Field-Effect Transistors

TEXAS INSTRUMENTS ELECTRONICS SERIES

The Engineering Staff of
Texas Instruments Incorporated ▪ TRANSISTOR CIRCUIT DESIGN

Runyan ▪ SILICON SEMICONDUCTOR TECHNOLOGY

Sevin ▪ FIELD-EFFECT TRANSISTORS

Field-Effect Transistors

Leonce J. Sevin, Jr.
Senior Engineer
Semiconductor Components Division
Texas Instruments Incorporated

McGRAW-HILL BOOK COMPANY
New York   San Francisco   Toronto   London   Sydney

Field-Effect Transistors

Copyright © 1965 by Texas Instruments Incorporated. All Rights Reserved. Printed in the United States of America. This book, or parts thereof, may not be reproduced in any form without permission of Texas Instruments Incorporated. *Library of Congress Catalog Card Number 64-8624*

Information contained in this book is believed to be accurate and reliable. However, responsibility is assumed neither for its use nor for any infringement of patents or rights of others which may result from its use. No license is granted by implication or otherwise under any patent or patent right of Texas Instruments or others.

56355

Preface

Although much literature on the subject of field-effect transistors has accumulated, no attempts have been made to present a consistent and comprehensive treatment of the theory, characterization, and applications in a single monograph. This book is an attempt to fill the void. The reader needs no other background material on field-effect transistors to become fairly proficient at designing electronic circuits using them.

The material presented is the lore accumulated during the author's tenure as an applications engineer at Texas Instruments Incorporated. Part of this time was devoted to the study and characterization of field-effect transistors and the development of circuits using them as active elements.

The level of presentation is aimed at practicing electronic circuit design engineers including those without engineering degrees who have attained the status of engineers through years of fruitful experience. Indeed, this material should be assimilated easily by senior engineering students.

Chapter 1 presents the physical behavior of the field-effect transistor. We hope that the method of presentation will give the circuit designer an insight into field-effect behavior and give him confidence to apply the device in new situations.

Chapter 2 contains discussion and development, where applicable, of the electrical characteristics of field effects important in circuit applications.

No claim of originality is made for any of the theoretical development in Chapters 1 and 2. The arbitrary charge distribution technique for analyzing field-effect behavior must be credited to William Shockley and R. R. Bockemuehl, and partly at least to James Clerk Maxwell.

Chapters 3, 4, and 5 contain the development of the field-effect transistor as a circuit element in low-level linear, nonlinear, and power circuits respectively. The material in Chapter 5 is mostly conjecture since there are no really high-power field effects yet available.

Chapter 6 contains additional circuit examples that were not necessary to the developments presented in the preceding three chapters. Chapter 7 is a discussion of the field-effect structures applied to integrated-circuit techniques. Both unipolar and MOS-type field effects are discussed.

I am indebted to many people whose cooperation and assistance made this task immeasurably easier. Stanley W. Holcomb, engineer and free hand circuit designer par excellence who contributed not only directly but also indirectly through numerous discussions, arguments, and shared experiences. John R. Miller, technical publications manager at Texas Instruments, provided encouragement and much constructive criticism in preparing the manuscript.

My gratitude also is due to Dr. J. R. Biard for many valuable hours of interesting and lively discussion and consultation; to J. A. Walston and Paul Vergez, and to my former associates, Bob Crawford, Ralph Dean, Vern Glover, and Lorimer Hill.

The manuscript was typed by Mrs. Dorcas Helms. This young lady, working from the incredible neohieroglyphics that pass for the author's handwriting, produced a very professional manuscript.

Leonce J. Sevin, Jr.

Contents

Preface v

Chapter 1. Theory of the Unipolar Field-effect Transistor 1

    1.1 Behavior of a Reverse-biased Diode 1
    1.2 Conductance of a Semiconductor Bar 3
    1.3 Semiconductor Bar With PN Junctions 4
    1.4 Channel Current Flow 7
    1.5 Behavior of the FET at Pinch-off and Beyond 11
    1.6 Certain Figures of Merit and the Importance of the Transfer Characteristic 17
    1.7 Construction of Double-diffused Field-effect Transistors . 24
    1.8 Surface Field-effect Transistors 24
          BIBLIOGRAPHY 29

Chapter 2. FET Characteristics 32

    2.1 Static Characteristics 32
    2.2 Dynamic Characteristics 41
          BIBLIOGRAPHY 50

Chapter 3. FET's in Low-level Linear Circuits 51

    3.1 Basic FET Amplifier Circuits 51
    3.2 Biasing FET Amplifiers 57
    3.3 Distortion in FET Amplifiers 59
    3.4 Wideband Audio and Video Amplifiers 61
    3.5 Direct-coupled Amplifiers 70
    3.6 FET's and High Fidelity 72
    3.7 Automatic Gain Control 76

Chapter 4. FET's in Nonlinear Circuits 80

    4.1 FET Squaring Circuits 80
    4.2 FET's in Switching Circuits 84
    4.3 FET Choppers and Commutators 89

Chapter 5. Blue Skies Dept.: The Power FET 93

    5.1   The Power FET and Power Transistor Compared 93
    5.2   Thermal Stability and Driver Requirements 94
    5.3   Hypothetical Power FET 94
    5.4   Practical Audio Amplifier 97
    5.5   Servo Amplifier 101

Chapter 6. Further Applications 102

    6.1   D-C Amplifiers 102
    6.2   A-C Amplifier 108
    6.3   Electronic Voltmeters 108
    6.4   Oscillators 111
    6.5   Sample-hold Circuit 113
    6.6   Bilateral Constant-current Source 114

Chapter 7. FET's in Integrated Circuits 116

    7.1   Unipolar FET's 118
    7.2   Surface FET's 123
          BIBLIOGRAPHY 125

    Index 127

Field-Effect Transistors

1

Theory of the Unipolar Field-effect Transistor

A field-effect transistor (FET) is essentially a semiconductor current path whose conductance is controlled by applying an electric field perpendicular to the current. The electric field results from reverse-biasing a PN junction.

1.1 BEHAVIOR OF A REVERSE-BIASED DIODE

When a PN junction is formed, mobile current carriers near either side of the junction diffuse across the junction and recombine with carriers of the opposite type, leaving equal and opposite electric charges on either side of the junction. This electric charge causes an electrostatic potential which tends to prohibit further diffusion of current carriers across the junction. The potential that completely prevents movement of holes and electrons across the junction is called the *barrier potential* or *contact potential*. The region on either side of the junction from which mobile carriers have disappeared, is called the *space-charge* or *depletion layer*. The behavior of the space-charge layer with reverse voltage is illustrated in Fig. 1.1. The impurity atom density is plotted vs. distance perpendicular to the junction. The areas shaded by diagonal lines are the space-charge layers caused by the junction contact potential ϕ. The dotted areas are the space-charge layers due to an externally applied reverse voltage on the PN junction. The reverse voltage is applied between two nonrectifying metal contacts attached to each end of a semiconductor bar containing a PN junction; this is shown in Fig. 1.2. The polarity of the external voltage, V_E, is in the same sense as the polarity of the contact potential, positive on the N side of the junction and negative on the P side. Thus, the effect of V_E is to increase the height of the space-charge layer in Fig. 1.1 to H_{V1}, where $H_{V1} = H_{VE} + H_\phi$. In Fig. 1.1, the donor impurity density on the N side is larger than the acceptor impurity density on the P side of the junction; therefore, H_{V1} must be larger than H'_{V1} by the same ratio to maintain equilibrium. This condition is implied by the relation $NH'_{V1} = PH_{V1}$. If the impurity density on one side of the junction is purposely made very large compared to the impurity density on the other side, then the space-charge-layer width on the more dense side can be

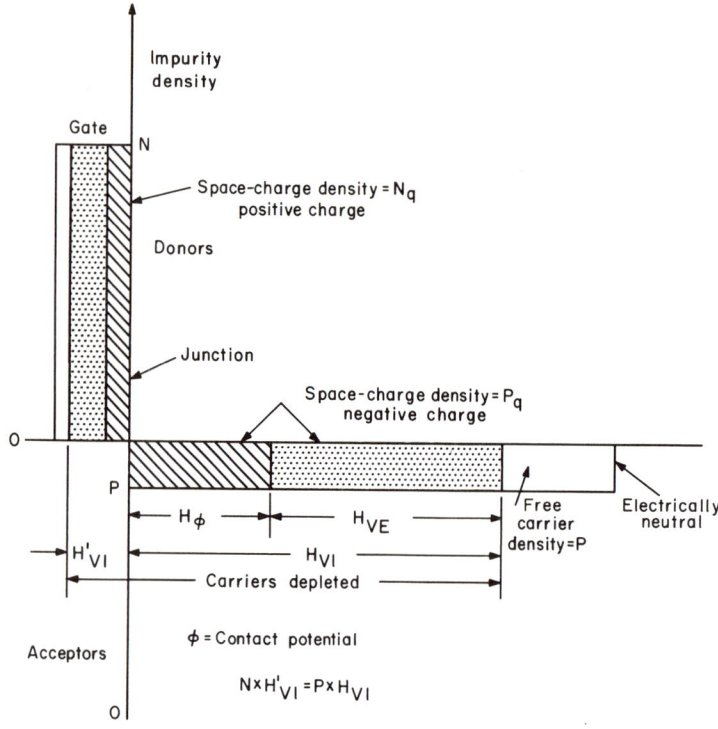

Fig. 1.1. Behavior of space-charge layer with reverse bias.

considered negligible compared to the width on the less dense side. The significance of this fact will become evident later in the discussion of field-effect operation.

When the external reverse voltage on a diode is increased, a current flows in the external circuit. Internally, this shows up as the movement of holes and electrons away from the junction. For every free electron that enters the external wire on the N side (see Fig. 1.2), a free electron leaves the external wire and recombines with a hole on the P side. This current flows until the electrostatic potential across the junction comes into equilibrium with the external battery. Thus, the current that flows is a displacement current and the PN junction behaves as a capacitor.

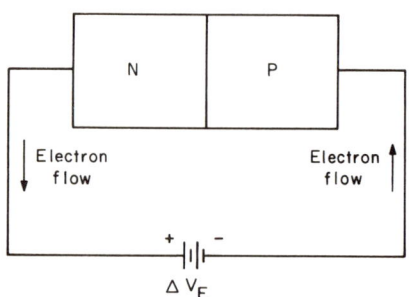

Fig. 1.2. Transient (capacitive) current in the external circuit of a reverse-biased diode.

The capacitance of the junction is equal to that of a parallel-plane capacitor having the same cross-sectional area as the PN junction, and a separation between the plates equal to the total height of the depletion layer, $H'_{V1} + H_{V1}$ in Fig. 1.1.

1.2 CONDUCTANCE OF A SEMICONDUCTOR BAR

Consider a semiconductor bar having the dimensions shown in Fig. 1.3 and with ohmic (nonrectifying) contacts represented by the shaded parts at each end. The conductance between the ohmic contacts depends on the dimensions of the bar and the conductivity. It is well known that a pure semiconductor, such as silicon or germanium, is an extremely poor conductor because few mobile carriers (holes and electrons) are available to conduct current—only those electron-hole pairs generated by thermal energy. However, if a large concentration of impurity atoms is introduced into the crystal (at any reasonable temperature, say above 200°K), it can be assumed that one current carrier is available per impurity atom. Further, if the impurity atom density is large compared to the density of thermally generated electron-hole pairs, then the current carriers available can be considered to be of only one type; e.g., if the impurity atoms are all trivalent (acceptors), only holes are available to carry current. The conductance of the bar is then proportional to the total number of carriers present.

If acceptor atoms are introduced into a bar such that the concentration is uniform in every part of the bar, the conductance of the bar is

$$G_b = \frac{q\mu WH}{L} p \qquad (1.1)$$

where q = electronic charge, 1.6019×10^{-19} coulomb
μ = carrier drift mobility, cm²/volt-sec
p = impurity density, atoms/cm³

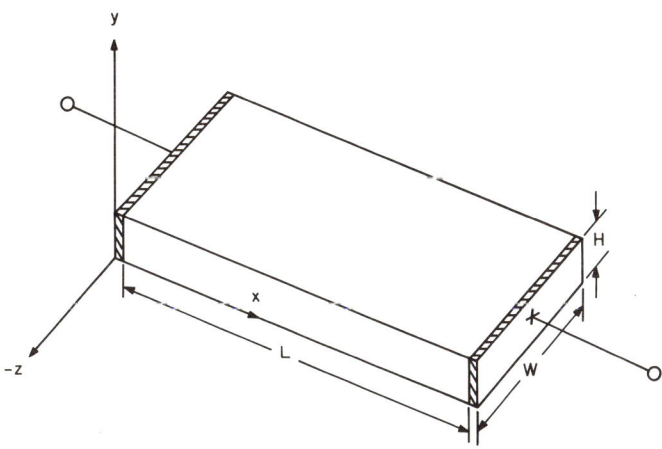

Fig. 1.3. A semiconductor bar.

4 Field-effect Transistors

Uniformly doped semiconductor crystals are grown from the pure molten semiconductor where the impurities are introduced in the melt. When impurities are introduced by some diffusion process, the concentration will then be a function of one or more of the dimensions of the bar in Fig. 1.3; if impurities are diffused uniformly over one of the surfaces, then the concentration will be a function of only the dimension normal to that surface. If in Fig. 1.3 impurities are diffused through the xz plane, then the conductance of the bar can be calculated from

$$G_b = q\mu \frac{W}{L} \int_0^H p(y)\, dy \qquad (1.2)$$

1.3 SEMICONDUCTOR BAR WITH PN JUNCTIONS

Figure 1.4 shows the bar of Fig. 1.3 with N-type impurities introduced into opposite sides, forming two PN junctions with the semiconductor bar. The two N regions are electrically connected, and a reverse voltage V_{GS} is applied to the two junctions. The interesting part of our semiconductor bar is now the region between the two junctions. If the impurity concentration in the N regions is purposely made very high compared to that in the bar, then the space-charge layer due to the contact potential and the external bias V_{GS} will extend almost entirely into the region of the bar between the junctions. Now, since there are virtually no free carriers in the space-charge layer (except those generated by heat), the conductance between the source and drain terminals in Fig. 1.4 is almost entirely determined by the region between the PN junctions not depleted of free carriers by the reverse junction voltage. Thus, it is easy to see how the applied voltage V_{GS} controls the conductance of our semiconductor bar.

Figure 1.4 is, in its essentials, a field-effect transistor. It is simplified to be sure, but essentially it is as we said, "a semiconductor current path whose conductance is controlled by applying an electric field perpendicular to the current." Shockley called this type of device, which uses the space-charge layer about a PN junction

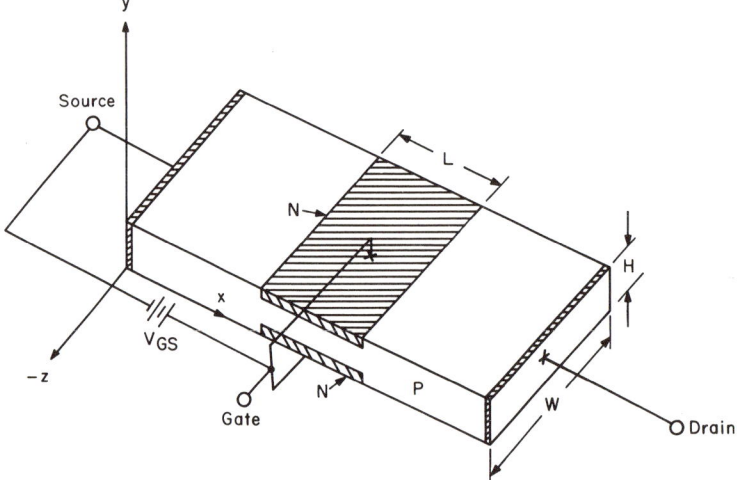

Fig. 1.4. Semiconductor bar with PN junctions.

to control conductance, a *unipolar field-effect transistor* because the transistor current flow is due to one type of carrier. The terminals of the FET in Fig. 1.4 are labeled drain, gate, and source, respectively, and are analogous to the plate, grid, and cathode, respectively, of a vacuum tube, and the collector, base, and emitter, respectively, of a junction or bipolar transistor. Note that the two N regions are electrically connected to form the gate.

It has probably been observed by now that the drain and source terminals are interchangeable. In the jargon of the art, the part of the bar directly between the PN junctions is called the *channel* and is the active part of the FET; the parts of the bar between the ends of the channel and the ohmic contacts can be considered lumped bulk resistors.

At this point let us define the terms we will apply to the channel dimensions throughout this book. Referring to Fig. 1.4, we shall call the dimension of the channel in the x direction its *length,* in the z direction its *width,* and in the y direction its *height.*

If an external voltage V_{DS} is applied between the drain and source terminals in Fig. 1.4, the resulting current that flows is obviously a function of V_{GS}, but what may not be obvious is that this current, the drain current, is not a linear function of V_{DS}. The flow of current in the channel produces an ohmic drop that affects the reverse bias on the junction, and hence affects the cross-sectional area of the conducting part of the channel. Thus, the resistance between the drain and source terminals is nonlinear.

The development of FET theory presented here is based on the approach taken by R. R. Bockemuehl, of the General Motors Research Laboratories, in his paper[1,*] published in the IEEE Transactions on Electron Devices of January, 1963. This paper has contributed much to the understanding of FET behavior; it is probably as significant in the advancement of the state of the art of field-effect transistors as the Ebers and Moll paper was to the junction transistor.

To simplify matters at the outset, we can take advantage of the symmetry in Fig. 1.4 and consider only the behavior of the space-charge layer on the channel side of one junction; of course, the fact that the example was initially chosen to be symmetrical helps; nevertheless, let us consider the simplified channel profile in Fig. 1.5. The polarities of the external voltages V_{SG} and V_{SD} are correct for a P-channel FET. The shaded regions are the space-charge layers due to ϕ, (H_ϕ), the reverse voltage due to V_{GS}, and the ohmic drop produced by I_D. In a P-channel FET, I_D is negative, so the drop it produces causes the highest reverse voltage at the drain end of the channel; hence, the peculiar wedge shape of the space-charge layer. The parallelepiped section of the channel shown in Fig. 1.5 represents the lower half of the region between the PN junctions of the silicon bar in Fig. 1.4.

In the following development, it is assumed that the space-charge density of any part of the space-charge layer is equal to the free-carrier density that would exist in the same place if the reverse bias on the junction were removed; it is also assumed that the space-charge-layer width on the gate side of the junction is negligible.

---

* Superscript numbers refer to bibliography entries at end of chapter.

6 Field-effect Transistors

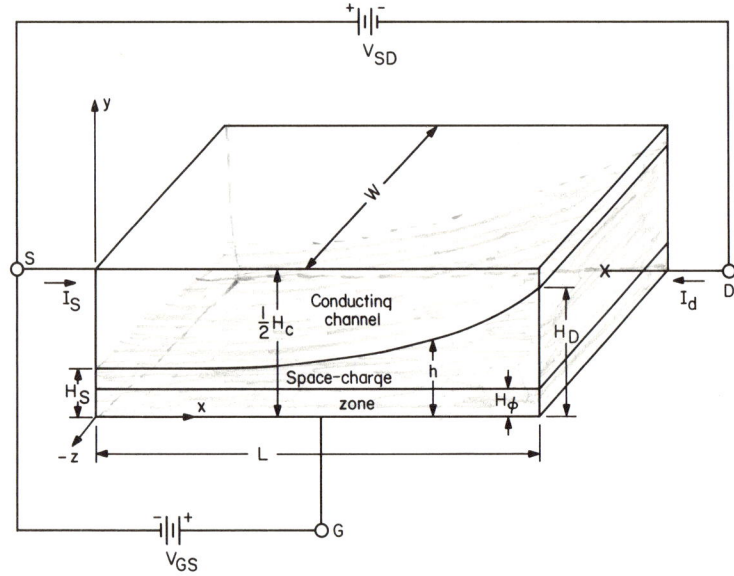

Fig. 1.5. The active channel.

In the region of fixed space charge, the behavior of the charge density with applied potential can be expressed by Poisson's equation:

$$\nabla^2 V = -\frac{\rho}{\epsilon} \qquad (1.3)$$

where ∇ = vector space operator $\left(1x\dfrac{\delta}{\delta x} + 1y\dfrac{\delta}{\delta y} + 1z\dfrac{\delta}{\delta z}\right)$

ρ = charge density, coulombs/cm^3 = qp

ϵ = dielectric constant, farads/cm

If we momentarily consider the behavior of the space-charge layer with V_{GS}, Eq. (1.3) becomes

$$\frac{d^2V}{dy^2} = -\frac{qp(y)}{\epsilon} \qquad (1.4)$$

Integrating Eq. (1.4) and substituting in the boundary condition that $dV/dy = 0$ when $y = h$, we have

$$\frac{dV}{dy} = -\frac{q}{\epsilon}\left[\int_0^y p(y)\,dy - \int_0^h p(y)\,dy\right] \qquad (1.5)$$

The integral $\int_0^y p(y)\,dy$ in Eq. (1.5) is the total number of impurity atoms in a section of the channel bounded by 0 and y having unit dimensions in the x and z directions. Therefore, the total charge in this part of the channel is

$$Q(y) = q\int_0^y p(y)\,dy \qquad (1.6)$$

To find the voltage across the space-charge layer, we integrate Eq. (1.5) once more and find that

$$V(h) - V_{GS} = \frac{q}{\epsilon}\int_0^h y p(y)\, dy \tag{1.7}$$

Equation (1.7) yields two significant and important bits of information about the FET. One is the capacitance per unit cross section of the gate-to-channel diode:

$$C = \frac{dQ}{dV(h)} = \frac{\epsilon}{h} \tag{1.8}$$

This is the capacitance of a parallel-plane diode of unit area having a plate separation of h, and in this case is only half the total per unit area capacitance, owing to symmetry.

The other important bit of information available from Eq. (1.7) can be seen when the upper limit of the integration is allowed to go to $1/2 H_c$. This yields the diode reverse voltage that removes all the free charge from the channel, which Shockley[2] called the "pinch-off" voltage: the current path has been pinched off; i.e., it no longer has the ability to conduct current because all the free carriers have been removed. In his definition, Shockley neglected the effect of the contact potential; for our purpose we will define the pinch-off voltage as the value of V_{GS}, the external gate-to-source voltage, that produces pinch-off:

$$V_P = \frac{q}{\epsilon}\int_{H_\phi}^{\frac{1}{2}H_c} y p(y)\, dy \tag{1.9}$$

Evaluation of the integral of Eq. (1.9) requires that $p(y)$ be known; e.g., if the gate-to-channel diode was formed by alloying a pentavalent metal to a silicon bar uniformly doped with acceptor impurities, the junction would be "abrupt," $p(y)$ could be assumed to be constant in the channel, and the pinch-off voltage would be

$$V_P = \frac{qP_o\left(\frac{1}{4}H_c^2 - H_\phi^2\right)}{2\epsilon} = \frac{qP_o H_c^2}{8\epsilon} - \phi \tag{1.10}$$

This is the pinch-off voltage derived by Shockley; the assumptions made were that $p(y) =$ a constant (P_o), and $\phi = 0$.

Figure 1.6 shows some arbitrary impurity density distributions. The impurity density must go through zero at the junction (by definition) and behave in some manner as a function of y; the functions shown are not necessarily practical, but serve to illustrate possible functions for $p(y)$. Free carriers cannot exist to the left of the dashed boundary of H_ϕ. Plots such as that shown in Fig. 1.6 are called *impurity profiles*.

1.4 CHANNEL CURRENT FLOW

The channel current density, according to Ohm's law, is

$$J_x = -\sigma_c \frac{dV}{dx} \tag{1.11}$$

8 Field-effect Transistors

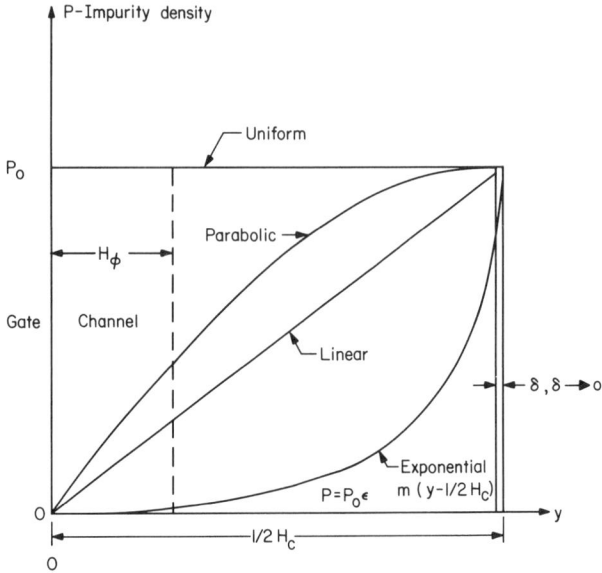

Fig. 1.6. Some arbitrary impurity profiles.

where σ_c = channel conductivity = $q\mu p$

Assuming no current flows in the space-charge zone, the drain current is

$$I_D = 2qW\mu \frac{dv}{dx} \int_h^{\frac{1}{2}H_c} p(y)\, dy \tag{1.12}$$

The factor 2 appears because of symmetry. The expression dv/dx is a function of the lower limit h in the integral, requiring that $h(x)$ be determined, which of course requires that $p(y)$ be known; thus, the generality of our development is in danger of being compromised. We can overcome this minor obstacle, however, by taking advantage of current continuity and dV/dy from differentiating Eq. (1.7). Multiplying both numerator and denominator of the right-hand side of Eq. (1.12) by dy we have

$$I_D\, dx = 2qW\mu \frac{hp(h)}{\epsilon} dh \int_h^{\frac{1}{2}H_c} p(y)\, dy \tag{1.13}$$

The integral in Eq. (1.13) can be written as

$$\frac{1}{q}\left[Q\left(\frac{1}{2}H_c\right) - Q(h)\right]$$

The expression in the brackets gives the total free charge in the parallelepiped bounded by $y = h$ and $y = \frac{1}{2}H_c$ having unit dimensions in the x and z directions.

When Eq. (1.13) is integrated from $x = 0$ to $x = L$, the corresponding limits on the right-hand side are from $y = H_S$ to $y = H_D$. Therefore,

$$I_D = \frac{2qW\mu}{\epsilon L} \int_{H_S}^{H_D} \left[Q\left(\frac{1}{2}H_c\right) - Q(h)\right] hp(h)\, dh \tag{1.14}$$

From Eq. (1.14) it is quite apparent that the current is a function of the space-charge-layer heights at both the source and drain ends of the channel (H_S and H_D respectively), which in turn are functions of the voltages across the space-charge layer at each end of the channel; i.e.,

$$H_S = H_\phi + F(V_{GS}) \qquad (1.15)$$

$$H_D = H_\phi + G(V_{DS} - V_{GS}) \qquad (1.16)$$

At this point it is convenient to define I_{DSS}, the drain current that flows when the external gate and source terminals are shorted together. By this definition, I_{DSS} is the drain current for any value of V_{DS}, but we will restrict the definition to the current that flows when V_{DS} has a value that causes H_D to equal $\frac{1}{2}H_c$; then

$$I_{DSS} = \frac{2qW\mu}{\epsilon L} \int_{H_\phi}^{\frac{1}{2}H_c} \left[Q\left(\frac{1}{2}H_c\right) - Q(h) \right] hp(h)\, dh \qquad (1.17)$$

Of course this is anomalous, since if $H_D = \frac{1}{2}H_c$, the conducting channel height is zero and no current can flow. Or what is more likely, if the conducting channel height goes to zero, then the current density must go to infinity, since we know intuitively that when $V_{GS} = 0$ we can not possibly cause I_{DSS} to go to zero by increasing V_{DS}. The answer to this dilemma is that H_D can not equal $\frac{1}{2}H_c$, because there is a fundamental limit on the current density in the conducting channel, and therefore a limit on how narrow the conducting channel can be. If the limiting value of current density is J_{max}, then $I_{DSS} = WJ_{(max)}\delta$ where δ is the narrowest possible conducting channel. Equation (1.17) is therefore only an approximation. The real upper limit on the integral is $(\frac{1}{2}H_c - \delta)$, but the approximation is a good one because $\frac{1}{2}H_c$ is much greater than δ. More will be said of this pinched-off behavior in Sec. 1.5.

The transconductance or mutual conductance of an FET, if we apply the familiar vacuum-tube definition, is

$$g_m = \frac{\delta I_D}{\delta V_{GS}} \qquad (1.18)$$

This can be found by applying the following relationship:

$$\frac{\delta I_D}{\delta V_{GS}} = \left(\frac{\delta I_D}{\delta H_S}\right)\left(\frac{\delta H_S}{\delta V_{GS}}\right) + \left(\frac{\delta I_D}{\delta H_D}\right)\left(\frac{\delta H_D}{\delta V_{GS}}\right) \qquad (1.19)$$

The partial derivatives $\delta I_D/\delta H_S$ and $\delta I_D/\delta H_D$ can be evaluated from Eq. (1.14); i.e.,

$$\frac{\delta I_D}{\delta H_S} = \frac{2qW\mu}{\epsilon L}\left[Q\left(\frac{1}{2}H_c\right) - Q(H_S)\right]H_S\, p(H_S) \qquad (1.20)$$

and

$$\frac{\delta I_D}{\delta H_D} = \frac{2qW\mu}{\epsilon L}\left[Q\left(\frac{1}{2}H_c\right) - Q(H_D)\right]H_D\, p(H_D) \qquad (1.21)$$

Now, from Eq. (1.17), the partial derivative of h, the space-charge-layer height at any x, with respect to the voltage across the space-charge layer $V(h)$ at that same x, is

$$\frac{\delta V(h)}{\delta h} = \frac{q}{\epsilon} h p(h) \tag{1.22}$$

Using Eq. (1.22) and the relationships in Eqs. (1.15) and (1.16), we have

$$\frac{\delta V_{GS}}{\delta H_S} = \frac{q H_S p(H_S)}{\epsilon} \tag{1.23}$$

and

$$\frac{\delta V_{GS}}{\delta H_D} = -\frac{q H_D p(H_D)}{\epsilon} \tag{1.24}$$

Substituting Eqs. (1.20), (1.21), (1.23), and (1.24) back into Eq. (1.19),

$$\frac{dI_D}{dV_{GS}} = g_m = \frac{2W\mu}{L} \Big[Q(H_D) - Q(H_S) \Big] \tag{1.25}$$

The implication of Eq. (1.25) is extremely interesting; it says that the transconductance of the FET is simply the conductance of the rectangular parallelepiped having dimensions $H_D - H_S$, W, and L (the factor 2 appears because of the symmetry in the example). At $I_D = I_{DSS}$, $H_D = \frac{1}{2}H_c$, and $H_S = H_\phi$, the transconductance is

$$g_{max} = \frac{2W\mu}{L} \Big[Q\Big(\frac{1}{2}H_c\Big) - Q(H_\phi) \Big] \tag{1.26}$$

This is the conductance of the channel at small-signal conditions, i.e., no external gate bias and only very small incremental values of V_{DS} about zero volts.

The output conductance can be derived in a similar manner:

$$g_D = \frac{\delta I_D}{\delta V_{DS}} = \Big(\frac{\delta I_D}{\delta H_S}\Big)\Big(\frac{\delta H_S}{\delta V_{DS}}\Big) + \Big(\frac{\delta I_D}{\delta H_D}\Big)\Big(\frac{\delta H_D}{\delta V_{DS}}\Big) \tag{1.27}$$

$$\frac{\delta H_S}{\delta V_{DS}} = 0 \tag{1.28}$$

$$\frac{\delta H_D}{\delta V_{DS}} = \frac{\epsilon}{q H_D p(H_D)} \tag{1.29}$$

Therefore,

$$g_D = \frac{2W\mu}{L} \Big[Q\Big(\frac{1}{2}H_c\Big) - Q(H_D) \Big] \tag{1.30}$$

Equation (1.30) is the conductance of the rectangular parallelepiped portion of the channel bounded by H_D and $\frac{1}{2}H_c$ in the y direction. According to Eqs. (1.17) and (1.30), when $I_D = I_{DSS}$, $g_D = 0$ and the drain terminal of the device behaves as a constant-current source; these results are shown graphically in Fig. 1.7. The channel is in effect a nonlinear resistor between the drain and source terminals. As V_{DS} is increased, H_D approaches $\frac{1}{2}H_c$ and g_D approaches zero. At $V_{DS} = V_P - V_{GS}$,

Theory of the Unipolar Field-effect Transistor 11

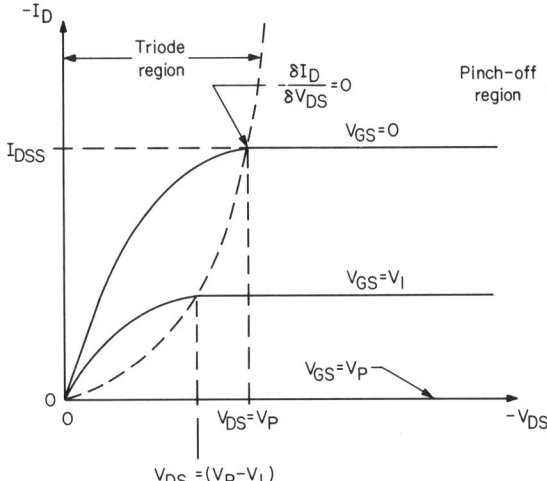

Fig. 1.7. Regions of operation.

g_D is zero, and further increase in V_{DS} produces no increase in I_D, because the upper limit on H_D is $\frac{1}{2}H_c$. This is only a first-order approximation, of course. When H_D approaches $\frac{1}{2}H_c$, other effects influence the behavior of the output characteristics in Fig. 1.7 such that g_D never gets to zero before the gate-to-drain diode goes into avalanche breakdown. These other effects will be discussed in Sec. 1.5.

The FET characteristics shown in Fig. 1.7 are commonly divided into two regions, the triode region and the pinch-off region. The two regions are separated by the dashed curve, which is the locus of pinch-off points, i.e., the points where $V_{DS} = V_P - V_{GS}$.

1.5 BEHAVIOR OF THE FET AT PINCH-OFF AND BEYOND

In Fig. 1.7 the triode region and the pinch-off region are separated by a locus of points at which $g_D = 0$. This condition results from the assumption that the height of the conducting channel goes to zero. To visualize why and how the assumption breaks down, and how much error is introduced by making the assumption, it is necessary to examine the height of the depletion layer in Fig. 1.5 as a function of x. We shall see that, because of the implicit nature of $h(x)$, it will be easier to derive and examine $x(h)$.

From Fig. 1.5 and Eq. (1.13), at a given constant pinch-off drain current I_{DP} and any value of h,

$$dx = \frac{2qW\mu}{I_{DP}\epsilon} hp(h)\, dh \left[Q\left(\frac{1}{2}H_c\right) - Q(h) \right] \quad (1.31)$$

Integrating and substituting the boundary condition at $x = 0$, $h = H_S$, and normalizing x to $x = L$, we have

$$\frac{x}{L} = \frac{2qW\mu}{I_{DP}\epsilon L} \int_{H_s}^{h} \left[Q\left(\frac{1}{2}H_c\right) - (Qh) \right] hp(h)\, dh \quad (1.32)$$

The integral in the above equation can be divided into two parts by realizing that

$$\int_{H_S}^{h} F(h)\,dh = \int_{H_S}^{\frac{1}{2}H_c} F(h)\,dh - \int_{h}^{\frac{1}{2}H_c} F(h)\,dh \tag{1.33}$$

where $F(h) = [Q(\frac{1}{2}H_c) - Q(h)]hp(h)$

Substituting Eq. (1.33) into Eq. (1.32), we have

$$\frac{x}{L} = \frac{2qW\mu}{I_{DP}\epsilon L}\left\{\int_{H_S}^{\frac{1}{2}H_c} F(h)\,dh - \int_{h}^{\frac{1}{2}H_c} F(h)\,dh\right\} \tag{1.34}$$

It is readily apparent that the first half of the right-hand side of Eq. (1.34) is equal to I_{DP}; therefore,

$$\frac{x}{L} = 1 - \frac{2qW\mu}{I_{DP}\epsilon L}\int_{h}^{\frac{1}{2}H_c}\left[Q\left(\frac{1}{2}H_c\right) - Q(h)\right]hp(h)\,dh \tag{1.35}$$

To a first approximation, I_{DP} depends only on the lower limit of the integral at $x/L = 0$. This means simply that since the channel is pinched off at the drain end and the drain current is no longer affected by the drain voltage, the current can only be controlled by the gate-to-source voltage. Therefore, for any fixed value of the gate-to-source voltage, I_{DP} is constant, and the remainder of the expression involving the integral, i.e.,

$$\frac{2qW\mu}{\epsilon L}\int_{h}^{\frac{1}{2}H_c}\left[Q\left(\frac{1}{2}H_c\right) - Q(h)\right]hp(h)\,dh$$

has the value I_{DP} at $x/L = 0$ and $h = H_S$, and vanishes as h approaches $\frac{1}{2}H_c$ at $x/L = 1$. To plot x/L as a function of h, the function $p(h)$ must be known; much insight into FET behavior can be gained from such a graph, so we will choose an arbitrary function for $p(h)$ and plot Eq. (1.35). The function chosen is the easiest and most obvious; i.e., $p(h) = $ a constant $= P_o$. A uniform distribution can be closely approximated in practice by alloying the gate to a uniformly doped channel. When substituting $p(h) = P_o$ in Eq. (1.35) we must remember that part of the channel charge is depleted by H_ϕ or

$$Q(h) = \int_{H_\phi}^{h} P_o\,dh = P_o(h - H_\phi) \tag{1.36}$$

when all the substitutions are made, we have

$$\frac{x}{L} = 1 - \frac{I_o}{I_{DP}}\left[1 - 3\left(\frac{h}{\frac{1}{2}H_c}\right)^2 + 2\left(\frac{h}{\frac{1}{2}H_c}\right)^3\right] \tag{1.37}$$

where $I_o = \dfrac{2qW\mu P_o(\frac{1}{2}H_c)^3}{6\epsilon L}$

for a uniformly doped channel. I_o is the current that would flow in the channel if there were no contact potential, or if enough forward bias were applied between the gate and the channel to reduce H_ϕ to zero. The latter is impossible, of course, since it would require an infinite forward current in the gate-to-channel diode. Equation (1.37) is plotted in Fig. 1.8 (with axes rotated 90°); I_{DP} is the running parameter. I_{DP} is equal to I_{DSS} when the gate is shorted ($V_{GS} = 0$) and H_ϕ is the

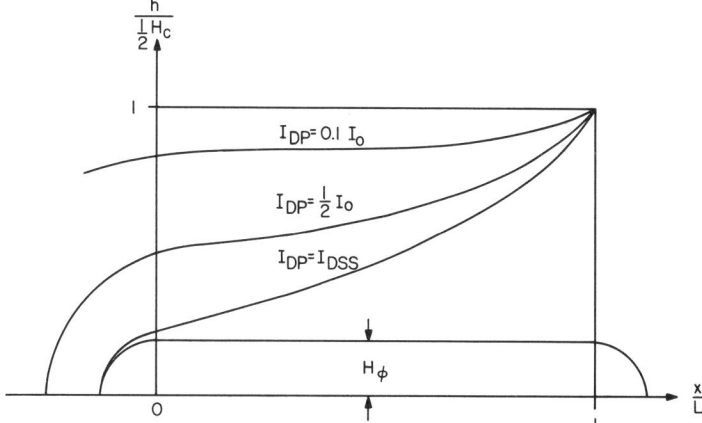

Fig. 1.8. Behavior of the space-charge layer with pinch-off drain current.

depletion-layer height at the source end. If the bulk resistance of the semiconductor between $x/L = 0$ and the source contact is neglected, the height of the depletion layer at $x/L = 0$ depends only on the external voltage between the gate and source terminals. The entire channel can be depleted, reducing I_{DP} to zero when $V_{GS} = V_P$. This same voltage is necessary between the source and drain terminals when $V_{GS} = 0$ (again neglecting bulk resistances outside the channel) to produce pinch-off, i.e., to cause $h = \frac{1}{2}H_c$ at $x/L = 1$.

Now, as has been suggested previously, the conducting channel is never completely pinched-off; i.e., h cannot quite equal $\frac{1}{2}H_c$ because this would imply infinite current density, hence, infinite electric field in the direction of current flow. For high values of electric field in semiconductors, voltage-current relationships no longer conform to Ohm's law; but rather, charge velocities begin to saturate and ultimately reach a value which becomes independent of the applied electric field until the semiconductor goes into avalanche breakdown. The range of values of electric field in silicon at room temperature is from 1 kilovolt/cm, where Ohm's law begins to fail, to about 60 kilovolts/cm at avalanche breakdown. These extremely high values of electric field are seldom reached in ordinary lumped resistors (at least not more than once per resistor). There is a relatively small body of literature based on the study of this high field phenomenon. One of the more readily understood reports is that of J. B. Gunn;[3] Fig. 1.9 shows Gunn's data on the variation of drift velocity with electric field in a sample of N-type germanium. Other reports [4,5] are available that show similar curves for N-and P-type silicon. This graph shows that carrier mobility is, in general, not constant and can be assumed to be constant only at fairly low values of electric field (below 100 volts/cm in germanium). The drift velocity saturates at an applied field of about 3 kilovolts/cm above which the mobility must vary as approximately one over the applied electric field. No simple relationship between the coordinates of Fig. 1.9 has been published, although it seems that an exponential relationship might fit the data. However, even if a simple exponential is substituted for μ in Eq. (1.12), things get quite unmanageable. For example, the simple expression

14   Field-effect Transistors

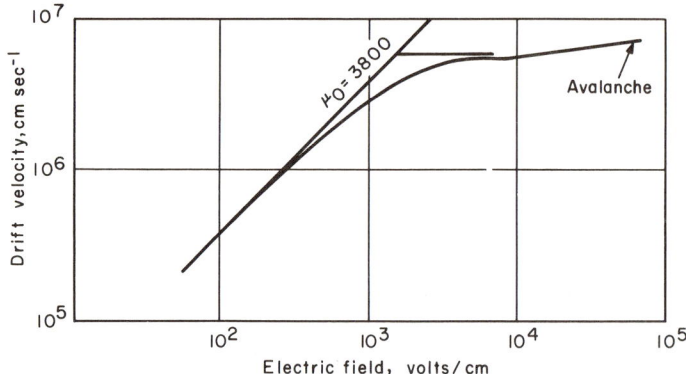

Fig. 1.9. Effect of high electric fields on carrier velocity.

$$\mu = \mu_o \epsilon^{-E/E_o}$$

can be made to fit the data quite well where E_o is the value of electric field when the drift velocity is 63.2 per cent of its saturation value. Substitution of this value of μ into Eq (1.12) yields

$$I_{DP} = 2qW\left[Q\left(\tfrac{1}{2}H_c\right) - Q(h)\right]\mu_o \frac{dV}{dx}\epsilon^{-(1/E_o)dV/dx} \qquad (1.38)$$

An attempt to integrate both sides of this equation yields a weird form that the author has not seen before. The integration of Eq. (1.38) being beyond the scope of this discussion (not to mention the author), we will confine ourselves to making some qualitative observations about velocity-limited current flow in the pinched-off channel.

First, a modification of the charge depletion model in Fig. 1.8 is dictated by the carrier velocity limit. In the plot of $h(x)$ of Fig. 1.8, values of x/L outside the range $0 \leq x/L \leq 1$ are not shown; we can get a clearer picture of the effects of velocity limiting by modifying Fig. 1.8 to include the space-charge layers in the bulk semiconductor outside the active channel. Figure 1.10 shows this modification with a semiconductor bar with a "front" and a "back" gate; we have reverted to showing both halves of the channel that we started out with in Fig. 1.4. In Fig. 1.10, V_{DS} is imagined to be held constant at a value just large enough to pinch the channel off (cause velocity limiting) at the drain end; if the resistive drop in the bulk semiconductor beyond $x = L$ is assumed to be negligible, the narrowest part of the conducting channel will be at approximately $x = L$. V_{GS} is shown to be variable and is to be varied from $V_{GS} = 0$ to $V_{GS} = V_P$. When $V_{GS} = 0$, $I_{DP} = I_{DSS}$; at the source end of the channel the depletion-layer height is H_ϕ, hence the active channel height is $H_c - 2H_\phi$. The channel narrows to $H_c - \delta$ at the drain end.

Figure 1.9 shows that the carrier velocity limit does not occur abruptly, but undergoes a gradual transition from the Ohm's law region to constant velocity. Let us, for the sake of simplifying the argument, assume that the velocity limit does occur abruptly and that the behavior is that of the asymptotes drawn onto the curve in Fig. 1.9. Then when the channel narrows to δ, the current density has reached its limiting value and further increase in V_{DS} can cause no further increase in current. The radius of the space-charge layer about a point at $x = L, y = 0$, is $\frac{1}{2}H_c$. Again, this assumes no IR drop beyond $x = L$. Increasing V_{GS} in the direction of reverse bias narrows the conducting channel over its entire length; the example in Fig. 1.10 implies enough change in V_{GS} to reduce I_{DP} to $\frac{1}{2}I_{DSS}$. At $x = 0$, the space-charge-layer height is now some value greater than H_ϕ. At $x = L$, it takes less current for the channel IR drop to produce pinch off; velocity limiting will occur at a lower I_{DP}, therefore, δ must decrease. Clearly, δ is a function of V_{GS}. When V_{GS} is increased to V_P, the active channel current must go to zero, thus δ must also vanish. Only bulk leakage currents flow in the depletion layer, the channel is uniformly pinched off, and the space-charge layer radii about $x = 0$ and $x = L$ are equal.

Now let us look at Fig. 1.11 and see what happens at constant V_{GS} are variable V_{DS} beyond pinch-off. Assuming that velocity limiting occurs abruptly, once the limit is reached further increase in V_{GS} can not increase I_{DP}, so δ must be unaffected by V_{DS} beyond pinch-off. Increasing V_{DS} simply lengthens the path of velocity-limited flow (Δx in Fig. 1.11) because of the further widening (Δh) of the space-charge layer. It is interesting to note that once the critical value of electric field dV/dx is reached, the field does not increase in proportion to $V_{DS} - V_P$. To illustrate this, let us assume that

$$\frac{dV}{dx} \approx \frac{\Delta V}{\Delta x} = \frac{V_{DS} - V_P}{\Delta x} \tag{1.39}$$

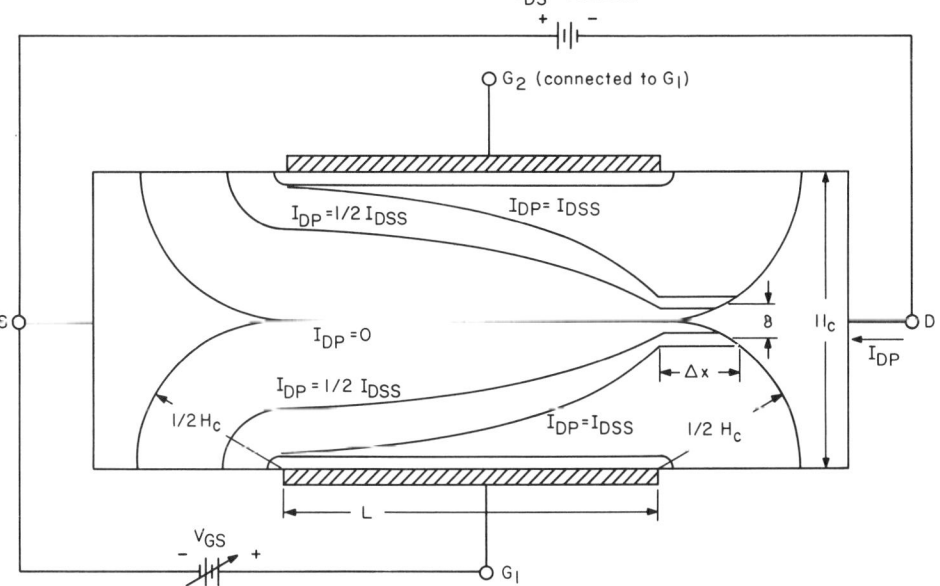

Fig. 1.10. Effect of V_{GS} on δ and Δx.

16 Field-effect Transistors

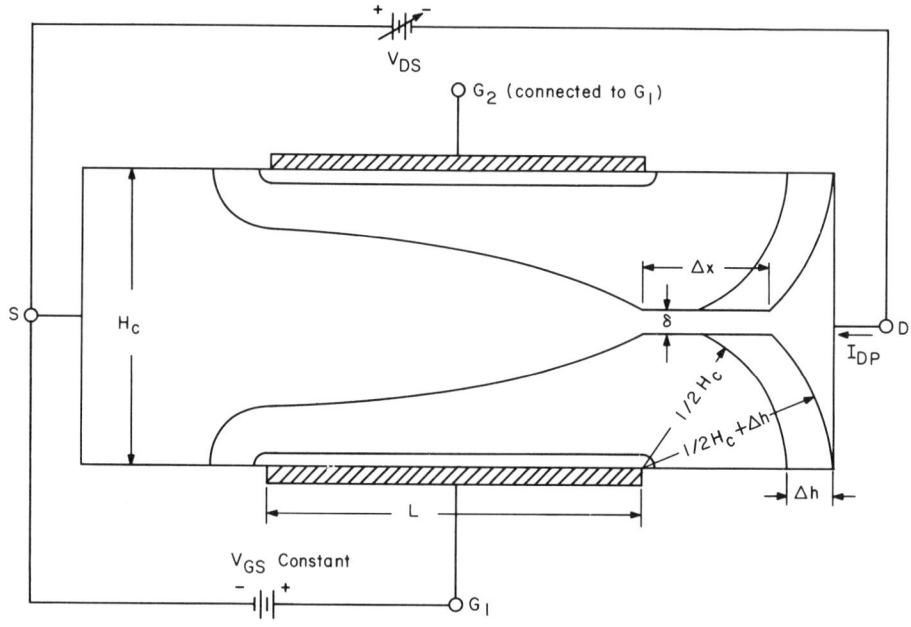

Fig. 1.11. Effect of V_{DS} on Δx.

If we assume further in Fig. 1.11 that δ is very small compared to $\frac{1}{2}H_c$, then we can write

$$\Delta x = \left(\frac{1}{2}H_c + \Delta h\right)^2 - \left(\frac{1}{2}H_c\right)^2 \qquad (1.40)$$

where
$$V\left(\frac{1}{2}H_c + \Delta h\right) = V_{DS} \qquad V_{DS} \geq V_P$$

$$V\left(\frac{1}{2}H_c\right) = V_P$$

From Eq. (1.7), remembering that the examples in Figs. 1.8, 1.10, and 1.11 are assumed to be uniformly doped,

$$\left(\frac{1}{2}H_c + \Delta h\right)^2 = \frac{2\epsilon V_{DS}}{P_o} \qquad (1.41)$$

$$\left(\frac{1}{2}H_c\right)^2 = \frac{2\epsilon V_P}{P_o} \qquad (1.42)$$

then
$$\Delta x = \frac{2\epsilon}{P_o}(V_{DS} - V_P) = \frac{2\epsilon}{P_o}\Delta V$$

and finally
$$\frac{\Delta V}{\Delta x} = \frac{P_o}{2\epsilon} \qquad (1.43)$$

The significance of Eq. (1.43) is not its value, which is obviously incorrect because of the assumptions made, but that it is independent of V_{DS} once the critical field

Theory of the Unipolar Field-effect Transistor   17

is reached. This is interesting. Breakdown cannot be achieved by the drift field, dV/dx, in the channel, but breakdown will occur because of the voltage across the space-charge layers, i.e., the reverse voltage across the gate-to-channel diode.

Returning to reality and the fact that velocity limiting does not occur abruptly, we may expect that the "knee" of the output voltage-current characteristic of an FET will closely resemble the one in Fig. 1.9 and that the dynamic output resistance will increase, approaching open circuit (the curve to have always negative concavity), until breakdown occurs. High-resistance shunt paths through the space-charge layer from drain to source (inversion layers) can occur to limit the output resistance, a fact which can be used as an indicator of the quality of the manufacturing process.

1.6 CERTAIN FIGURES OF MERIT AND THE IMPORTANCE OF THE TRANSFER CHARACTERISTIC

Having seen the derivation of the important equations governing the behavior of an FET, we are now in a position to derive certain useful and interesting figures of merit. We can greatly simplify our task by taking advantage of the high output impedance beyond pinch-off. The reasoning is that, since drain current changes little once the channel is pinched off, the relationships that hold at pinch-off should be good approximations to the behavior of the FET beyond pinch-off, or in the so-called "pinch-off region" of Fig. 1.7. One must choose this route as an alternative to coming to grips with that windmill in Eq. (1.38).

Let us review the important relationships we have derived by restating them here:

Pinch-off Voltage:

$$V_P = \frac{q}{\epsilon} \int_{H_\phi}^{\frac{1}{2}H_c} y p(y)\, dy \tag{1.9}$$

Gate Capacitance per Unit Area:

$$C = \frac{\epsilon}{h} \tag{1.8}$$

Drain Current:

$$I_D = \frac{2qW\mu}{\epsilon L} \int_{H_S}^{H_D} \left[Q\left(\frac{1}{2}H_c\right) - Q(h) \right] h p(h)\, dh \tag{1.14}$$

Saturation Drain Current:

$$I_{DSS} = \frac{2qW\mu}{\epsilon L} \int_{H_\phi}^{\frac{1}{2}H_c} \left[Q\left(\frac{1}{2}H_c\right) - Q(h) \right] h p(h)\, dh \tag{1.17}$$

Transconductance:

$$g_m = \frac{2W\mu}{L} \left[Q(H_D) - Q(H_S) \right] \tag{1.25}$$

Zero Bias Transconductance:

$$g_{max} = \frac{2W\mu}{L}\left[Q\left(\frac{1}{2}H_c\right) - Q(H_\phi)\right] \quad (1.26)$$

In the pinch-off region, $\tfrac{1}{2}H_c$ is substituted for H_D in Eqs. (1.14) and (1.25). The capacitance in Eq. (1.8) was derived assuming no drain current flowing when I_D is not zero; h becomes a function of x, and C must be evaluated by integration, specifically:

$$C_G = 2\epsilon W \int_0^L \frac{dx}{h(x)} \quad (1.44)$$

where C_G is the total gate capacitance and is no longer expressed per unit area.

The figures of merit we are interested in writing are

$$M_1 = \frac{g_m}{C_G}$$

and

$$M_2 = \frac{g_{max}V_P}{I_{DSS}}$$

M_1 determines the maximum frequency of operation of an FET. The gate capacitance must charge through the channel resistance in series with the external resistances in the source and gate terminals. If these external resistances are reduced to zero, M_1 is the charging time constant, since, as we have previously determined, the channel resistance for any bias condition is $1/g_m$. The significance of M_2 is easily seen with the aid of Fig. 1.12, which is a general transfer curve of a P-channel FET ($I_D = F[V_{GS}]$, V_{DS} = constant). The slope of this curve at $V_{GS} = 0$ is g_{max}. If this slope is extended as a tangent until it intersects the V_{GS}

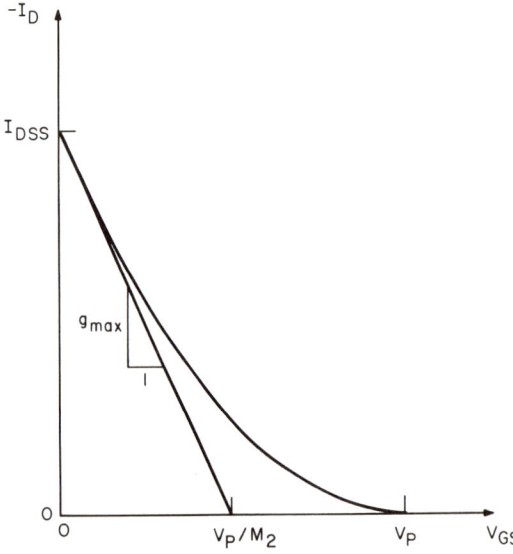

Fig. 1.12. The transfer curve.

axis, the point of intersection is V_{GS}/M_2. This is quickly proved from the similar triangles

$$g_{max} = M_2 \frac{I_{DSS}}{V_P} \qquad (1.45)$$

M_2 therefore is a measure of the nonlinearity of the forward transfer characteristic of an FET.

M_1 can be evaluated from Eqs. (1.25) and (1.44):

$$M_1 = \frac{g_m}{C_G} = \frac{\mu[Q(\tfrac{1}{2}H_c) - Q(H_S)]}{\epsilon \int_0^L \frac{dx}{h(x)}} \qquad (1.46)$$

Since we have seen that $h(x)$ is an implicit function, it is desirable to change the variable of integration from x to h; from Eq. (1.13):

$$dx = \frac{2qW\mu}{I_{DP}\epsilon}\left[Q\left(\frac{1}{2}H_c\right) - Q(h)\right]hp(h)\,dh \qquad (1.47)$$

Substituting Eq. (1.47) into Eq. (1.46), we get

$$M_1 = \frac{I_{DP}[Q(\tfrac{1}{2}H_c) - Q(H_S)]}{2qW \int_{H_S}^{\tfrac{1}{2}H_c} [Q(\tfrac{1}{2}H_c) - Q(h)]p(h)\,dh} \qquad (1.48)$$

Still another change of variable can be effected in the integral in the denominator, remembering that

$$qp(h) = \frac{dQ(h)}{dh}$$

When this is substituted into the integral, the limits change to $Q(\tfrac{1}{2}H_c)$ and $Q(H_S)$, and the integration can be performed:

$$\frac{1}{q}\int_{Q(H_S)}^{Q(\tfrac{1}{2}H_c)}\left[Q\left(\frac{1}{2}H_c\right) - Q(h)\right]dQ(h) = \frac{1}{2q}\left[Q\left(\frac{1}{2}H_c\right) - Q(H_S)\right]^2 \qquad (1.49)$$

Substituting this expression back into Eq. (1.48), we see that

$$M_1 = \frac{I_{DP}}{W[Q(\tfrac{1}{2}H_c) - Q(H_S)]} \qquad (1.50)$$

The denominator in the above equation represents all the free or mobile charge that would exist in one-half the channel between $y = \tfrac{1}{2}H_c$ and $y = H_S$ if the depletion-layer height were constant along the channel length at $h = H_S$. As is evident from Fig. 1.13, this is a fairly good approximation of the total charge "in transit" in the active channel, the area in the shaded rectangle being approximately equal to that of the undepleted part of the channel; it could not possibly be any less than half that area no matter what form $h(x)$ might take. Equation (1.50) has the form

20   Field-effect Transistors

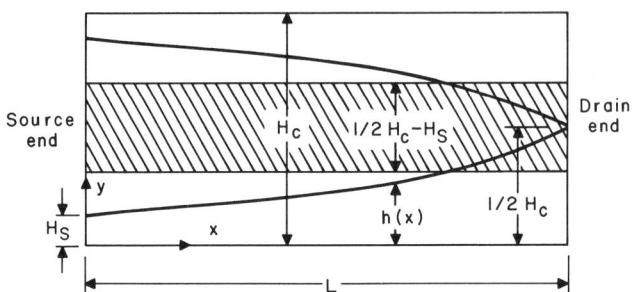

Fig. 1.13. Comparison C_G/g_m to transit time.

$$M_1 = \frac{I_{DP}}{\Delta Q_c} \tag{1.51}$$

and is the reciprocal of the transit time of mobile charges in the channel. Thus, we find that the cutoff frequency of an FET calculated from the effects of gate capacitance and channel resistance is in general agreement with the cutoff frequency we would calculate from transit time considerations. When used in untuned amplifiers, the FET's gate capacitance and external resistances will determine practicable cutoff frequencies.

To evaluate M_2 we need Eqs. (1.26), (1.9), and (1.17):

$$M_2 = \frac{[Q(\tfrac{1}{2}H_c) - Q(H_\phi)] \int_{H_\phi}^{\tfrac{1}{2}H_c} hp(h)\,dh}{\int_{H_\phi}^{\tfrac{1}{2}H_c} [Q(\tfrac{1}{2}H_c) - Q(h)]hp(h)\,dh} \tag{1.52}$$

A numerical solution requires knowledge of $p(h)$; i.e., the nonlinearity of the forward transfer curve depends on the channel impurity profile. Now let us refer to Fig. 1.6. These are impurity profiles, all of which are not necessarily practical, but they show that the net impurity density goes through zero at the junction boundary ($y = 0$), and increases in some manner on either side (from zero, you can only go up). In one case, the transition from N-type impurities to P-type impurities is abrupt and the impurity density is uniform in the channel at $p(y) = P_o$. This distribution, as we have said, is approximated by alloying the gate to a uniformly doped bar of semiconductor. The exponential distribution can be used as a tool to give any degree of nonlinearity by varying m from zero to a value approaching infinity. As m approaches infinity, all the charge appears concentrated at $y = \tfrac{1}{2}H_c$. Now we have two extremes of behavior of the impurity density; on the one hand, $p(y)$ jumps to P_o at $y = 0$ and remains constant for all values of y to the right of $y = 0$; and on the other hand, $p(y)$ is zero at $y = 0$ and remains so until y approaches $\tfrac{1}{2}H_c$, whereupon $p(y)$ abruptly jumps to P_o. We might call this a *spike profile*. These two extreme profiles when substituted into Eq. (1.52) will give the limits on M_2. These limits are

Uniform distribution, $M_2 = -3$
Spike profile, $M_2 = -2$

The effect of the contact potential is to alter the profiles in Fig. 1.6. After all, it is the free carriers associated with the impurity atoms that must be removed from the channel by the combination of the channel current and externally applied gate bias; this means that the lower limits on the integrals in Eq. (1.52) should be changed from zero to H_ϕ. This is a significant consideration for any FET whose channel is very narrow. The *free-carrier profile* approaches the spike profile case, regardless of the *impurity profile* (i.e., manufacturing process).

It is easily shown that a transfer curve like that in Fig. 1.12 that has $M_2 = 2$, is a parabola. We can demonstrate this in another fashion by evaluating Eq. (1.14) for both limiting $p(y)$; when this is done, we find that

$$I_{DP} = I_{DSS}\left[1 - 3\left(\frac{V_{GS}}{V_P}\right) + 2\left(\frac{V_{GS}}{V_P}\right)^{3/2}\right] \tag{1.53}$$

for the uniform channel, and

$$I_{DP} = I_{DSS}\left(1 - \frac{V_{GS}}{V_P}\right)^2 \tag{1.54}$$

for the spike channel.

Equation (1.53) was originally derived by Shockley,[2] and Eq. (1.54) is indeed a parabola. When Eqs. (1.53) and (1.54) are plotted on the same graph, they show that there is a surprisingly narrow range of possible transfer characteristics as indicated by the shaded region in Fig. 1.14.

Now Eq. (1.53) does not take the contact potential into account, but is only valid for FET's in which $\tfrac{1}{2}H_c$ is very large compared to H_ϕ. For any other case, the transfer curve must fall between the two shown in Fig. 1.14. The point here is that Eq. (1.54), with its delightful simplicity, is not a bad approximation of the transfer curve of any FET and is quite an accurate approximation for FET's with narrow channels (low pinch-off voltages) no matter how they are manufactured.

Consider the example of a silicon alloy FET with a pinch-off voltage of three times the contact potential ϕ. ϕ is about the same in FET's as in junction transistors, around 0.6 volt. Substituting $V_P = 3\phi$ into Eq. (1.10), we find that

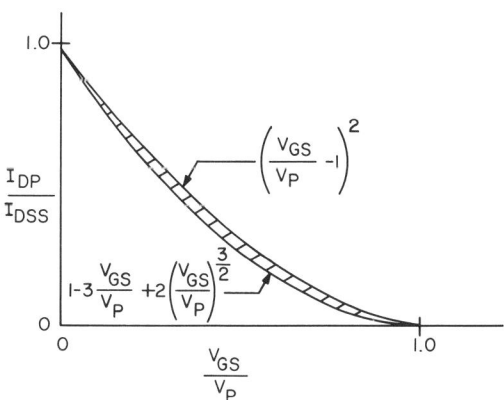

Fig. 1.14. Limits on transfer curve.

$$H_\phi = \frac{1}{2}\left(\frac{1}{2}H_c\right)$$

When this value of H_ϕ and $p(h) = p(y) = P_o$ are substituted into Eq. (1.52), the "initial slope" M_2 of the transfer curve for this example is $M_2 = 2.25$. Notice that even though the pinch-off voltage is three times the contact potential, M_2 is much closer to two than to three. This suggests that even for moderately small channels, the transfer curve converges rapidly to the parabolic or "square-law" case. Let us substitute our uniform distribution and integration limits into the current Eq. (1.14), and compare the result to the square-law curve. In this case, we will want an I_{DP} that changes the upper integration limit in Eq. (1.14) from H_D to $\frac{1}{2}H_c$, therefore,

$$I_{DP} = \frac{2qW\mu}{\epsilon L} P_o^2 \int_{H_S}^{\frac{1}{2}H_c} \left(\frac{1}{2}H_c - h\right) h\, dh \qquad (1.55)$$

After integrating and making the proper substitutions for H_S and $\frac{1}{2}H_c$:

$$I_{DP} = \frac{2qW\mu P_o^2(\frac{1}{2}H_c)^3}{6\epsilon L}\left[1 - 3\frac{V_{GS} + \phi}{V_P + \phi} + 2\left(\frac{V_{GS} + \phi}{V_P + \phi}\right)^{3/2}\right] \qquad (1.56)$$

By substituting $H_S = \frac{1}{4}H_c$ in Eq. (1.55), we find that

$$I_{DSS} = \frac{qW\mu P_o^2(\frac{1}{2}H_c)^3}{6\epsilon L} \qquad (1.57)$$

Substituting Eq. (1.57) into Eq. (1.56):

$$I_{DP} = 2I_{DSS}\left[1 - 3\frac{V_{GS} + \phi}{V_P + \phi} + 2\left(\frac{V_{GS} + \phi}{V_P + \phi}\right)^{3/2}\right] \qquad (1.58)$$

The graph of Eq. (1.58) compared to the square law is given by Fig. 1.15.

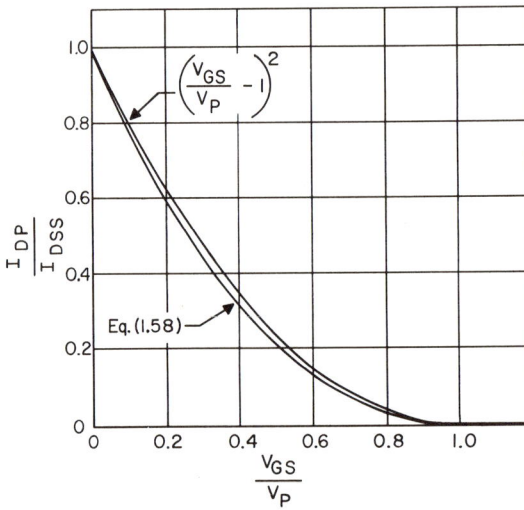

Fig. 1.15. Comparison of Eq. (1.58) to the square law.

Table 1.1

Unit no.	M_2	Manufacturer and type	Process
1	2.06	TI2N2497	Si double-diffused
2	2.15	TI2N2497	Si double-diffused
3	1.94	TI2N2497	Si double-diffused
4	2.01	TI2N2386	Si double-diffused
5	2.12	TI2N2386	Si double-diffused
6	2.11	Amelco FG37	Si diffused gate
7	1.92	Amelco FG36	Si diffused gate
8	2.05	Amelco FG36	Si diffused gate
9	2.10	TI2N3328	Si double-diffused, small geometry
10	1.83	TI2N3328	Si double-diffused, small geometry

When an FET is made by some diffusion process (as most PN junction FET's are nowadays), the free-carrier profile must fall somewhere between the uniform profile and the spike profile, and hence approach the square law even closer. (Ideal diffusion profiles are Gaussian in nature, of the form $1 - e^{-y^2}$, which for any reasonable argument approaches one.)

Because of its simplicity, the square-law approximation to the FET's transfer characteristic can be a very powerful engineering tool when the FET is used as a circuit element. More evidence of the accuracy of the approximation is available in Table 1.1, which shows values of M_2 determined graphically for ten FET's of various manufacture. Henceforth we will use the square-law approximation wherever the forward transfer characteristic is applicable. The values of I_{DSS} and V_P are necessary to completely describe the curve, and should be given by a manufacturer's data sheet. However, there is a good deal of uncertainty in measuring V_P directly because the curve changes so slowly at very low drain currents. A simpler and more accurate measurement of V_P can be obtained by using the figure of merit M_2 to solve for V_P from measured values of I_{DSS} and g_{max}, namely,

$$V_P = \frac{M_2 I_{DSS}}{g_{max}} \approx \frac{2 I_{DSS}}{g_{max}} \quad (1.59)$$

The convenience of measurement is enhanced by the fact that g_{max} and I_{DSS} are measured at the same bias point. If we use a test circuit like the one in Fig. 1.16, the two quantities can be measured simultaneously.

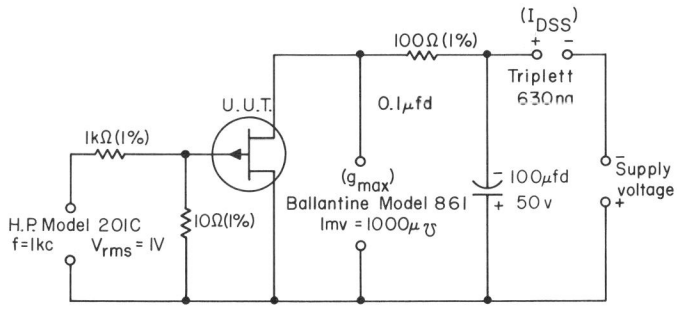

Fig. 1.16. Circuit for measuring g_{max} and I_{DSS}.

24 Field-effect Transistors

1.7 CONSTRUCTION OF DOUBLE-DIFFUSED FIELD-EFFECT TRANSISTORS

Figure 1.17 shows the construction and details of geometry of a double-diffused silicon P-channel FET. The "starting material" is a lightly doped N-type silicon slice into which two successive diffusions are made. The first is a P-type diffusion of somewhat higher initial concentration than the original uniform N-type doping, and the second diffusion (N type) is of much higher concentration than the first diffusion. The diffusion profiles are sketched in Fig. 1.18. The space-charge regions in Fig. 1.17 indicate that gate 1 exercises greater control over the active channel because of the heavier doping; gate 1 and 2 are electrically connected by extending the gate 1 diffusion outside the channel diffusion.

1.8 SURFACE FIELD-EFFECT TRANSISTORS

Some types of surface FET's have been envisioned and written about for many years. The idea predates even the point contact transistor. Variations of surface FET's are known by various other names, such as Induced Channel, Insulated Gate and MOS (metal-oxide-semiconductor) FET's.

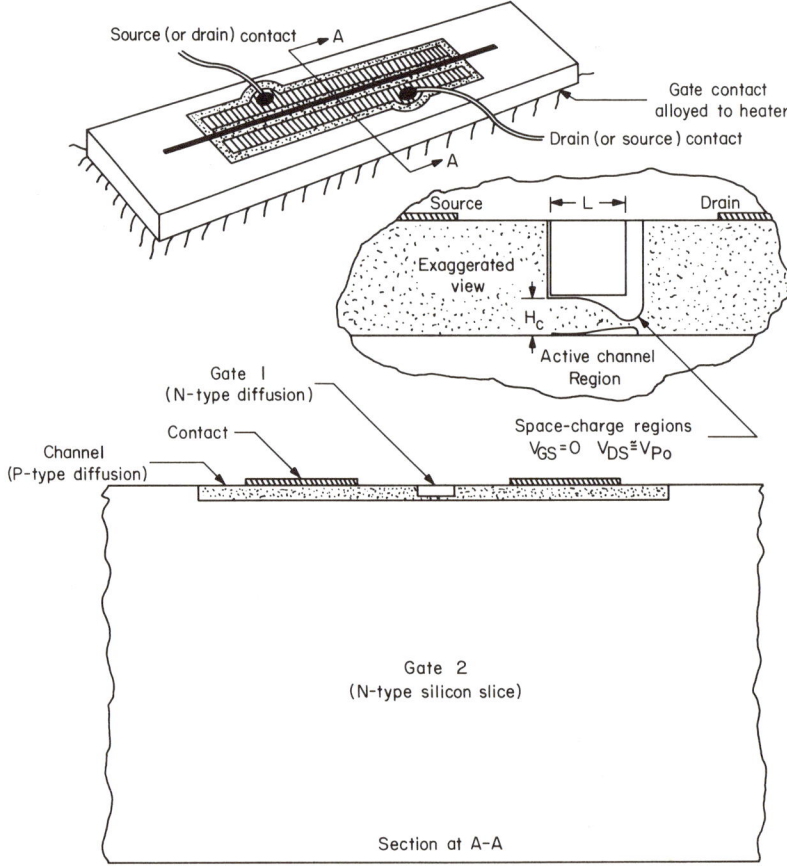

Fig. 1.17. Planar FET construction.

Theory of the Unipolar Field-effect Transistor

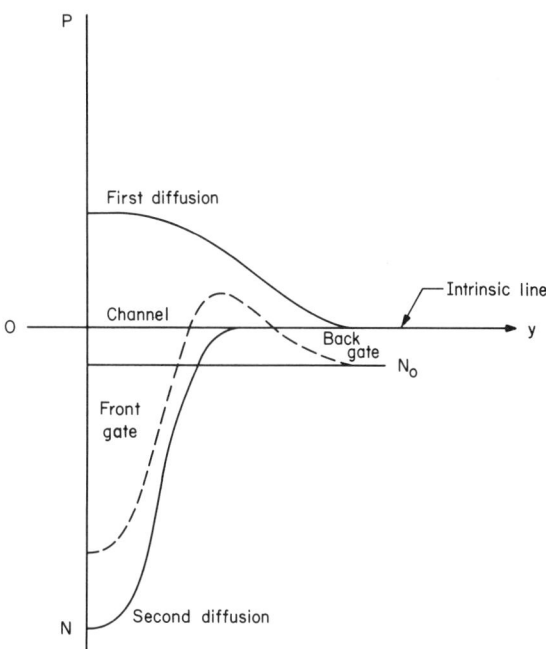

Fig. 1.18. Planar FET diffusion profile.

Such a device has the general construction shown in Fig. 1.19. Contact is made at two points on an intrinsic or lightly doped semiconductor through heavily doped regions of opposite type. In our example, the contacts are N type, and the substrate (or bar) is very lightly doped P type. The two N regions form the source and drain contacts, and a narrow, lightly doped N layer may or may not be initially diffused into the P-type bar between the source and drain contacts. This region between the two heavily doped N layers serves as the channel. The gate is a metal plate separated from the channel by an insulating dielectric. If an N layer is diffused into the channel region, the FET is said to have an *initial channel*. This initial channel can be depleted by the application of a negative voltage to the gate with respect to the channel, or by a current flowing from drain to source (conventional current) which makes the drain end more positive than the gate, just as in the case of the unipolar FET example of Fig. 1.5. As the channel near the drain end becomes almost completely depleted of free carriers, the drain current saturates and the dynamic output impedance becomes very high.

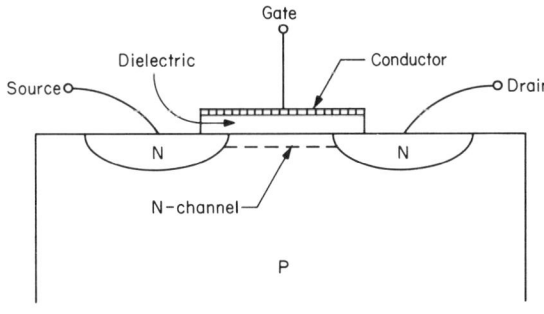

Fig. 1.19. Geometry of a surface FET.

26   Field-effect Transistors

The analysis of the unipolar FET's in the preceding sections can be applied to this device as well; a look at Fig. 1.19 will show that the surface FET closely approximates the spike profile distribution discussed in Sec. 1.6. As a matter of fact, if a surface FET's source bulk resistance is small, it is an excellent square-law device. When the surface FET of Fig. 1.19 is operated with negative voltage on the gate, it is said to be operating in the *depletion mode*.

The gate-to-channel contact of the unipolar FET is a PN junction. Given a small amount of forward bias, the gate begins to conduct heavily, greatly reducing the input impedance of the device as well as its power gain. Unlike that of the unipolar FET, the surface FET's gate is isolated from the channel through a dielectric, whose breakdown voltage is presumably the same for either polarity of applied voltage. Now when the gate of the device in Fig. 1.19 is biased with a positive voltage, a momentary drift field in the bulk semiconductor sweeps electrons toward the surface into the channel, increasing its current-carrying capacity. Even though the bulk semiconductor is doped with P-type impurities, at any instant there are always a number of electrons present that have been generated by thermal energy. This operation is known as the *enhancement mode*. A set of transfer characteristics and a set of output characteristics for a hypothetical surface FET are given in Figs. 1.20a and b, respectively; both depletion and enhancement mode are shown.

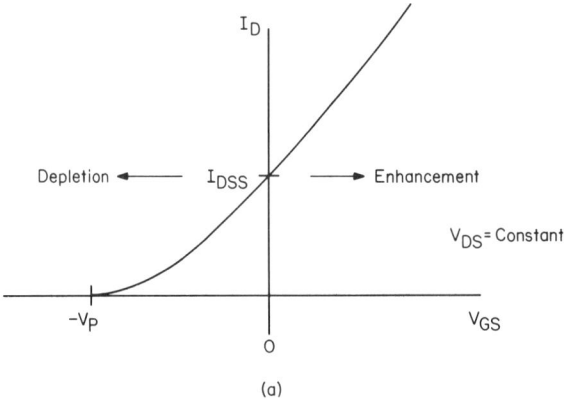

(a)

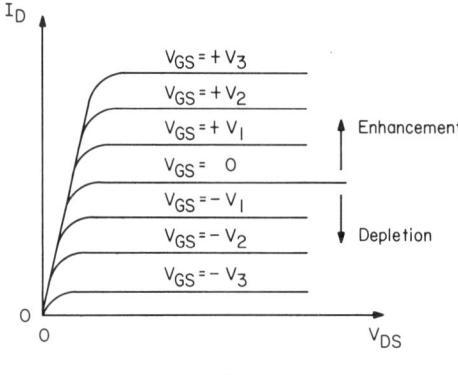

(b)

Fig. 1.20. Characteristics of a surface FET.

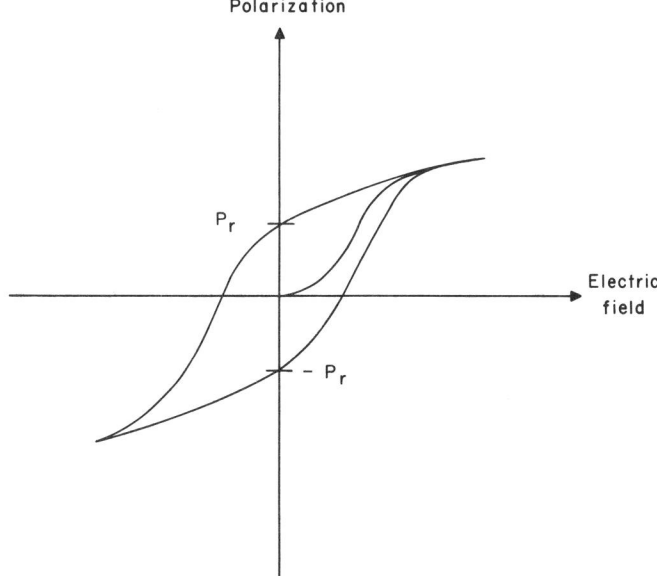

Fig. 1.21. Dielectric hysteresis loop.

For those interested in pursuing the theory of surface FET's in more detail, an excellent treatment by S. R. Hofstein and F. P. Heiman of RCA can be found in the IEEE Proceedings, September, 1963.

Early developmental models of surface FET's were plagued with defects (as one might expect). Most of the electric field in the dielectric terminated on fixed recombination centers rather than on mobile carriers, drastically reducing the ability of the gate voltage to control channel current. However, with advances in technology, this problem has virtually been eliminated. There remains one problem area with the device, and it may turn out to be a blessing in disguise; the problem is that dielectric materials available thus far exhibit ferroelectric properties. Figure 1.21 shows the hysteresis curve of a ferroelectric crystal; note the similarity to a magnetic hysteresis loop. A domain orientation theory[6] is used to explain this effect; the crystal is divided into many domains, each possessing a dipole moment. Presumably, these domains are initially randomly oriented. When an electric field is applied, the domains tend to align themselves with the field, and when it is removed, the domains do not all return to their initial orientation. This causes the remanent polarization P_r in Fig. 1.21. Materials that exhibit these properties are called *electrets* (after magnets).

The hysteresis of the electret itself offers no serious obstacle to the operation of a surface FET; *but* just as the characteristics of a magnet can be changed by heating it in the presence of a magnetic field, so can the characteristics of the electret be changed by heating it in the presence of an electric field. The characteristics can also be changed by shining light on the dielectric in the presence of an electric field. A built-in electric field in the dielectric of a surface FET causes part of the initial channel to be depleted (or enhanced) and decreases (or increases) I_{DSS}. Now in a circuit this poses a serious problem, because the FET will generally be

28 Field-effect Transistors

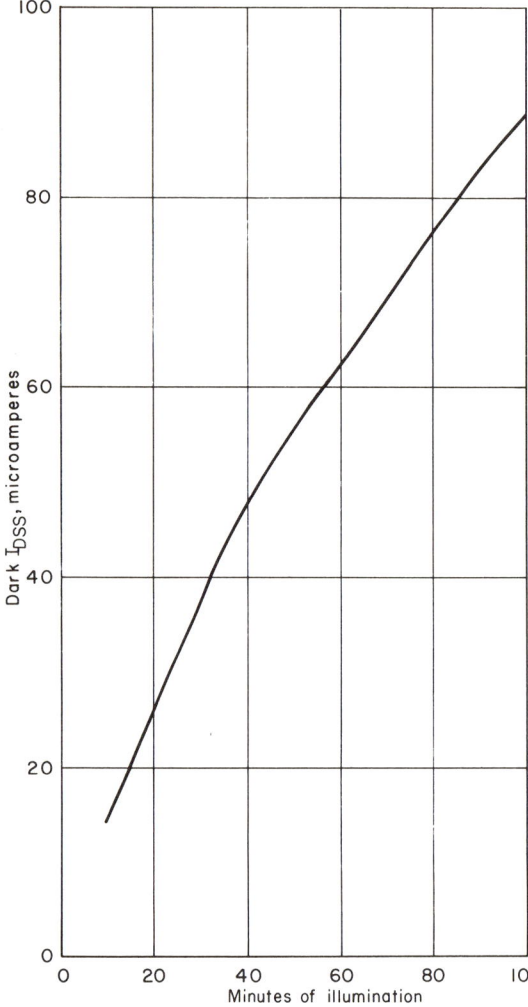

Fig. 1.22. Adaptation with light and positive gate bias.

operated with some bias, either depletion or enhancement, and if operated at an elevated temperature the characteristics will continuously change the value of I_{DSS} roughly proportional to the time integral of the applied field.

I_{DSS} will decrease with time in the presence of light or heat if a depletion bias is applied. This suggests that the device may be applied. It also suggests that the device may ultimately be useful as an adaptive element, i.e., an amplifying device whose gain has a memory. The adaption characteristics that have been observed on a particular surface FET are shown in Fig. 1.22. This curve was obtained by illuminating the dielectric with a $+1$-volt bias applied to the gate; every nine minutes or so the light was momentarily turned off while I_{DSS} was measured.

A large body of mathematics has been built up which describes adaptive computers, so called "learning machines" or "self-organizing machines," but there has been no large-scale machine development because no inexpensive adaptive element is available—none that can readily be applied to integrated circuits or some other high-density packing technique. This surface FET may be the answer to the problem. It is easily applicable to integrated-circuit techniques. But only sketchy adaption data are thus far available, indicating much work yet to be done. Nonetheless, the prospects are exciting.

BIBLIOGRAPHY

1. Bockemuehl, R. R.: Analysis of Field-effect Transistors with Arbitrary Charge Distribution, *IEEE Trans.*, Vol. ED-10, pp. 31–34, January, 1963.
2. Shockley, W.: A Unipolar Field-effect Transistor, *Proc. IRE,* vol. 40, pp. 1365–1376, November, 1952.
3. Cunn, J. B.: The Field Dependence of Electron Mobility in Germanium, *J. Electronics and Control,* vol. 2, p. 87, 1956.
4. Ryder, E. J., and W. Shockley: Mobilities of Electrons in High Electric Fields, *Phys. Rev.*, vol. 81, no. 1, pp. 139–140, January 1, 1951.
5. Ryder, E. J.: Mobility of Holes and Electrons in High Electric Fields, *Phys. Rev.*, vol. 90, no. 5, pp. 766–769, June 1, 1953.
6. Jona, F., and G. Shirane: "Ferroelectric Crystals," The Macmillan Company, New York, 1962.

General References

Bechtel, N. G.: A Circuit and Noise Model of the Field-effect Transistors, *Solid State Circuits Conference*, p. 92, 1963.
Bruncke, W. C.: Noise Measurement in Field-effect Transistors, *Proc. IEEE, Correspondence,* vol. 51, p. 378, February, 1963.
Burns, R. C., and E. D. Crawfis: Try Field-effect Transistors for Redundancy, *Electronic Design*, vol. 11, p. 74, January 18, 1963.
Cohen, J. M.: Generating Linear Waveforms with Field-effect Transistors, *Electronic Design,* vol. 11, p. 66, January 4, 1963.
Dacey, G. C., and I. M. Ross: The Field-effect Transistor, *Bell System Tech. J.,* vol. 34, pp. 1149–1189, November, 1955.
Dacey, G. C., and I. M. Ross: Unipolar "Field-effect" Transistor, *Proc. IRE,* vol. 41, p. 970, August, 1953.
Doucette, E. I.: Some Circuit Applications of the Field-effect Current Limiter, *Solid State Circuits Conference,* pp. 54–55, 1959.
Editorial Staff: Design with Field-effect Transistors Now?, *Electronic Equip. Eng.,* vol. 11, p. 68, March, 1963.
Editorial Staff: Field-effect Differential Millivoltmeter, *Electronic Design,* vol. 11, p. 66, August 2, 1963.
Editorial Staff: Field-effect Transistors, *solid/state/design,* vol. 4, p. 12, January, 1963.
Editorial Staff: Low-noise Preamplifier for Piezo Electric Transducers, *Electronic Design,* vol. 11, p. 92, August 16, 1963.
Editorial Staff: 1-MEV Test Shows FET's Resistant to Electron Irradiation, *Elec. Design News,* vol. 8, p. 6, May, 1963.
Etter, P. J., and B. L. H. Wilson: Inductance from a Field-effect Tetrode, *Proc. IRE, Correspondence,* vol. 50, p. 1828, August, 1962.
Evans, A. D.: Analyzing High-input-impedance FET Amplifiers, *Electronic Equip. Eng.,* vol. 11, p. 72, March, 1963.
Evans, A. D.: Characteristics of Unipolar Field-effect Transistors, *Electronic Industries,* vol. 22, p. 99, March, 1963.
Gerdes, R.: Field-effect Integrator Generates Long, Linear Sweeps, *Electronic Design,* vol. 11, p. 58, August 30, 1963.
Glover, L. L.: Using a New Device: Field-effect Transistor Oscillators, *Electronics,* vol. 36, p. 44, December 21, 1962.

Grosvalet, I., C. Motsch, and R. Tribes: Physical Phenomenon Responsible for Saturation Current in Field-effect Devices, *Solid-state Electronics,* Pergamon Press, vol 6, pp. 67–69, 1963.

Highleyman, W. H.: An Analog Multiplier Using Two Field-effect Transistors, *IRE Trans.,* vol. CS-10, p. 320, September, 1962.

Hill, L. O., D. O. Pederson, and R. S. Pepper: Synthesis of Bistable Circuits, *IEEE Trans.,* vol. CT-10, p. 25, March, 1963.

Hoerni, J. A., and B. Weir: Conditions for a Temperature Compensated Silicon Field-effect Transistor, *Proc. IEEE, Correspondence,* vol. 51, p. 1058, July, 1963.

Hofstein, S. R., and F. P. Heiman: The Silicon Insulated Gate Field-effect Transistor, *RCA Report,* Air Force Contracts, AF19(609)8836 and AF19(604)8040. Also, *Proc. IEEE,* vol. 51, no. 9, pp. 1190–1202, September, 1963.

Holden, P.: Reliability of the Field-effect Transistor, *Electro-Technology,* vol. 27, p. 58, July, 1963.

Huang, C., M. Marshall, and B. H. White: Field-effect Transistor Applications, *AIEE Trans., Commun. and Electronics,* vol. 75, no. 16, part 1, p. 323, July, 1956.

Ihanatola, H. K. J.: Design Theory of a Surface Field-effect Transistor, *Stanford Res. Tech. Rept. 1661-1,* 1961.

Johnson, E. O., and A. Rose: Simple General Analysis of Amplifier Devices with Emitter, Control, and Collector Functions, *Proc. IRE,* vol. 47, pp. 407–418, March, 1959.

Kulp, B. A., J. P. Jones, and A. F. Vetter: Electron Radiation Damage in Unipolar Transistor Devices, *Proc. IRE, Correspondence,* vol. 49, p. 1437, September, 1961.

Latham, D. C., D. J. Hamilton, and F. A. Lindholm: Low-frequency Operation of Four-terminal Field-effect Transistors, *IEEE Trans.,* vol. ED-11, pp. 300–305, June, 1964.

Latham, D. C., D. J. Hamilton, and F. A. Lindholm: New Modes of Operation for Field-effect Devices, *Proc. IEEE, Correspondence,* vol. 51, p. 226, January, 1963.

Lauritzen, P. O., and O. Leistiko, Jr.: Field-effect Transistors as Low-noise Amplifiers, *Solid State Circuits Conference,* p. 62, 1962.

Loe, J. M.: Single-stage AGC Has 40-db Control Range, *Electronic Design,* vol. 11, p. 72, August 2, 1963.

McIntosh, R. B., Jr.: Variable Resistance FET Gives 75-db Gain Control, *Electronic Design,* vol. 11, p. 56, August 30, 1963.

Martin, A. B.: Circuit Applications of the Field-effect Transistor, *Semiconductor Products,* parts 1 and 2, pp. 33–39, February, 1962; pp. 30–36, March, 1962.

Matzen, W.: Semiconductor Single Crystal Circuit Development, *Texas Instruments Report,* Technical Documentary Report ASD-TDR-63-281, Air Force Contract AF33(616)-6600, March, 1963.

Middlebrook, R. D.: A Simple Derivation of Field-effect Transistor Characteristics, *Proc. IEEE, Correspondence,* vol. 51, p. 1146, August, 1963.

Murphree, F. J., and J. H. Hammond: Field-effect Transistors Give Ultra-linear LF Sawtooth, *Electronic Design,* vol. 11, p. 62, July 5, 1963.

Olsen, D. R.: Equivalent Circuit for a Field-effect Transistor, *Proc. IEEE, Correspondence,* vol. 51, p. 254, January, 1963.

Parmer, W. F.: Evaluating Breakdown Voltage Characteristics of Field-effect Transistors, *Electronic Equip. Eng.,* vol. 11, p. 55, May, 1963.

Parmer, W. F.: Four Ways to Pair Field-effect and Conventional Transistors, *Electronic Design,* vol. 10, p. 44, August 16, 1962.

Prim, R. C., and W. Shockley: Joining Solutions at the Pinch-off Point in "Field-effect" Transistor, *IRE Trans.,* vol. ED, p. 1, December, 1953.

Richer, I., and R. D. Middlebrook: Power-law Nature of Field-effect Transistor Experimental Characteristics, *Proc. IEEE, Correspondence,* vol. 51, p. 1145, August, 1963.

Root, C. D., and L. Vadasz: Design Calculations for MOS Field-effect Transistors, *IEEE Trans.,* vol. ED-11, pp. 294–299, June, 1964.

Sah, C. T.: Characteristics of Metal-oxide Semiconductor Transistors, *IEEE Trans.,* vol. ED-11, pp. 324–345, July, 1964.

Sah, C. T.: Theory of Low Frequency Generation Noise in Junction-gate Field-effect Transistors, *Proc. IEEE,* vol. 52, pp. 795–814, July, 1964.

Sevin, L. J.: A Simple Expression for the Transfer Characteristic of FET's, *Electronic Equip. Eng.,* vol. 11, p. 59, August, 1963.

Shipe, W. H.: Field-effect Transistors Boost Voltmeter Input Impedance, *Electronic Design,* vol. 11, p. 71, June 7, 1963.

Shockley, W.: A Unipolar Field-effect Transistor, *Proc. IRE,* vol. 40, p. 1365, November, 1952.

Stone, H. A., Jr.: The Field-effect Tetrode, *1959 IRE Convention Record,* part 3, pp. 3–17.

Stone, H. A., Jr., and R. M. Warner, Jr.: The Field-effect Tetrode, *Proc. IRE,* vol. 49, p. 1170, July, 1961.

van der Ziel, A.: Gate Noise in Field-effect Transistors at Moderately High Frequencies, *Proc. IRE,* vol. 51, p. 461, March, 1963.

van der Ziel, A.: Thermal Noise in Field-effect Transistors, *Proc. IRE,* vol. 50, p. 1808, August, 1962.

van der Ziel, A., and J. W. Ero: Small-signal, High-frequency Theory of Field-effect Transistors, *IEEE Trans.,* vol. ED-11, pp. 128–135, April, 1964.

Wanlass, F. M., and C. T. Sah: Nanowatt Logic Using Field-effect Metal-oxide Semiconductor Triodes, paper presented at the 1963 International Solid State Circuits Conference, Philadelphia, Pa.

Warner, R. M., Jr.: A New Passive Semiconductor Device, *1958 IRE National Convention Record,* part 3, pp. 43–48.

Warner, R. M., Jr., and F. R. Carlson: Current-regulating Circuit Handles Loads in Ampere Range, *Electronic Design,* vol. 11, p. 54, August 30, 1963.

Warner, R. M., Jr., W. H. Jackson, E. I. Doucette, and H. A. Stone, Jr.: A Semiconductor Current Limiter, *Proc. IRE,* vol. 47, p. 44, January, 1959.

Wegener, H. A. R.: The Cylindrical Field-effect Transistor, *IRE Trans.,* vol. ED-6, p. 442, October, 1959.

Weimer, P. K.: The TFT—A New Thin-film Transistor, *Proc. IRE,* vol. 50, pp. 1462–1469, 1962.

2

FET Characteristics

2.1 STATIC CHARACTERISTICS

We saw in Chap. 1 that a great deal of information about an FET can be obtained from the static values of the zero-to-gate-bias drain current I_{DSS} and the pinch-off voltage V_P. In circuit calculations when using FET's, these two quantities often give most of the information necessary to design a circuit. Other FET characteristics (except for breakdown voltages) can often be regarded as causing only second-order effects; for the most part, they can be ignored. In some cases it is necessary to know something about the other static characteristics of the FET: gate leakage current, breakdown voltages, and the static drain-to-source resistance.

The static drain-to-source resistance, $r_{D(on)}$ (measured at $V_{GS} = 0$ and $V_{DS} = 0$ and very-small-signal voltage between drain and source to measure the resulting signal current) is mentioned because of its importance in chopper and AGC circuits. However, it will be shown later in this chapter that $r_{D(on)}$ can easily be calculated when I_{DSS} and V_P are known.

I_{DSS} and V_P. The triple subscript notation describing the quantity I_{DSS} is defined by IEEE Standard 56IRE28.S1. The first subscript indicates the object terminal, in this case the drain; the second subscript indicates which terminal is common, in this case the source; and the third subscript gives the condition of the remaining terminal with respect to the common terminal—in this case the gate remains, and the S stands for short. So by the standard definitions, I_{DSS} is the drain current of a common-source FET whose gate is shorted to the source, which is how we defined it in Chap. 1.

The set of FET output characteristics in Fig. 2.1 shows the dependence of I_{DSS} on the drain-to-source voltage. The reader can judge for himself the validity of constant-current approximation in the pinch-off region. We will explain those peculiar breakdown characteristics later in this chapter.

Although I_{DSS} is not greatly dependent on the drain-to-source voltage, it is a very strong function of temperature, as Fig. 2.2 attests. The temperature dependence of I_{DSS} can quickly be derived from Eq. (1.17). Differentiating this equation with respect to temperature (T) we have

FET Characteristics

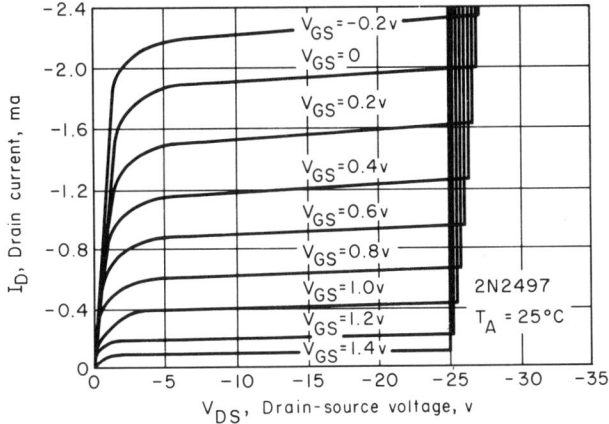

Fig. 2.1. Common-source drain characteristics.

$$\frac{dI_{DSS}}{dT} = \frac{2qW}{\epsilon L}\left(\frac{d\mu}{dT}\right)\int_{H_\phi}^{\frac{1}{2}H_c}\left[Q\left(\frac{1}{2}H_c\right) - Q(h)\right]hp(h)\,dh$$

$$-\frac{2qW\mu}{\epsilon L}\left[Q\left(\frac{1}{2}H_c\right) - Q(h)\right]H_\phi p(H_\phi)\frac{dH}{dT} \quad (2.1)$$

The two terms on the right-hand side of Eq. (2.1) can be reduced to much simpler forms. The first term can be reduced to $1/\mu(d\mu/dT)I_{DSS}$ by inspection, where $1/\mu(d\mu/dT)$ is by definition the temperature coefficient of the mobility. Simplification of the second term can be carried out in a straightforward manner by separating it into two parts. From Eq. (1.26):

$$\frac{2W\mu}{L}\left[Q\left(\frac{1}{2}H_c\right) - Q(h)\right] = g_{max} \quad (2.2)$$

When Eq. (1.9) is differentiated with respect to temperature, we find that the only

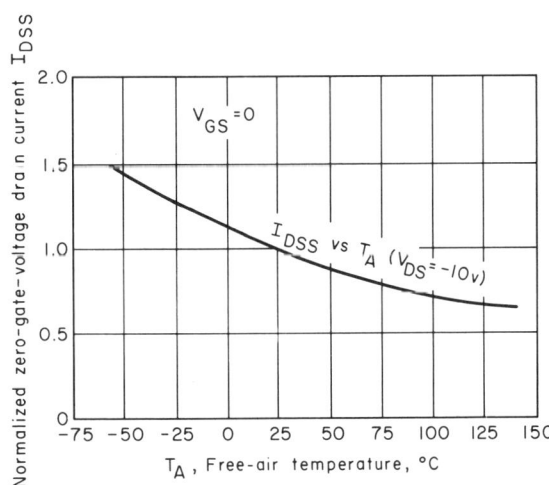

Fig. 2.2. Normalized zero-gate-voltage drain current and static drain-source resistance vs. free-air temperature.

34   Field-effect Transistors

temperature-dependent term is the lower limit of the integral (at temperatures above 200°K we can assume that all the carriers are activated and $p(y)$ is independent of temperature). The result of the differentiation is

$$\frac{dV_P}{dT} = -\frac{q}{\epsilon} H_\phi p(H_\phi) \frac{dH_\phi}{dT} \tag{2.3}$$

When these relationships are substituted into Eq. (2.1), a much simpler expression for dI_{DSS}/dT results:

$$\frac{dI_{DSS}}{dT} = \frac{1}{\mu} \frac{d\mu}{dT} I_{DSS} + g_{max} \frac{dV_P}{dT} \tag{2.4}$$

Further simplification is possible by making use of the figure of merit M_2 that we introduced in Chap. 1,

$$\frac{dI_{DSS}}{dT} = I_{DSS} \left(\frac{1}{\mu} \frac{d\mu}{dT} - \frac{M_2}{V_P} \frac{dV_P}{dT} \right) \tag{2.5}$$

Equation (2.5) implies a zero of dI_{DSS}/dT when

$$V_P = M_2 \mu \frac{dV_P/dT}{d\mu/dT} \tag{2.6}$$

Now this is a very significant and useful result. We now need to determine $d\mu/dT$ and dV_P/dT.

Mobility vs. temperature curves are given in several texts, the most familiar being Shockley's "Electrons and Holes in Semiconductors." The mobility vs. temperature curves given for silicon show that, over the temperature range of practical interest, a reasonable approximation to the mobility-temperature function is

$$\mu = \mu_o \left(\frac{T}{T_o} \right)^{-n} \tag{2.7}$$

where n depends on the impurity concentration. The term needed for substitution into Eq. (2.5) is the temperature coefficient of mobility, which we can now determine quite easily from Eq. (2.7):

$$\frac{1}{\mu} \frac{d\mu}{dT} = -\frac{n}{T} \tag{2.8}$$

Equation (1.7) is a general expression for the voltage across any arbitrary height of the channel space-charge zone; therefore we can write

$$\phi = \frac{q}{\epsilon} \int_0^{H_\phi} hp(h) \, dh \tag{2.9}$$

Differentiating this with respect to temperature we find that

$$\frac{d\phi}{dT} = \frac{q}{\epsilon} H_\phi p(H_\phi) \frac{dH_\phi}{dT} \tag{2.10}$$

Finally, from Eq. (2.3):

$$\frac{dV_P}{dT} = -\frac{d\phi}{dT} \quad (2.11)$$

Temperature dependence of the pinch-off voltage is entirely due to the variation of the contact potential. This is the same phenomenon that causes the dV_{BE}/dT in junction transistors. ϕ is a log function of the absolute temperature and is also a log function of the doping densities on either side of the PN junction. It is common practice, however, to use a linear approximation for the dV_{BE}/dT in transistors. A value of about ± 2.2 mv/C° (depending on whether the transistor is PNP or NPN) at low emitter current is satisfactory. This is probably familiar. Such an approximation will be adequate for our purposes, although it has been the author's experience that $d\phi/dT$ is more nearly -2.0 mv/C° in double-diffused FET's, probably because the gate and channel dopings are not quite the same as in the base and emitter of a transistor.

Returning to Eq. (2.6) and substituting $-n/T$ for $1/\mu(d\mu/dT)$, we see that the temperature coefficient of I_{DSS} will be zero on any device that fits all our assumptions, having a pinch-off voltage of

$$V_P = -\frac{M_2}{n} T \frac{d\phi}{dT} \quad (2.12)$$

Now we need a value for M_2/n. $M_2 = -2$ is a very good approximation for the silicon double-diffused FET's. We could use a value of n quoted several places in the literature, but these values range from 1.7 to 2.7 because the mobility-temperature function depends on the doping level in the sample being tested. We need the value of n for the doping levels used in the FET channel. This can be obtained from a plot of I_{DSS} vs. temperature for FET's with high pinch-off voltages; i.e., $d\phi/dT$ has negligible effect on dI_{DSS}/dT. This plot is shown in Fig. 2.3

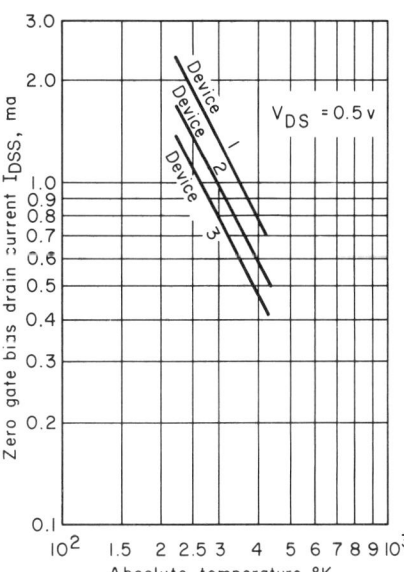

Fig. 2.3. I_{DSS} vs. temperature for FET's with high v_P.

36 Field-effect Transistors

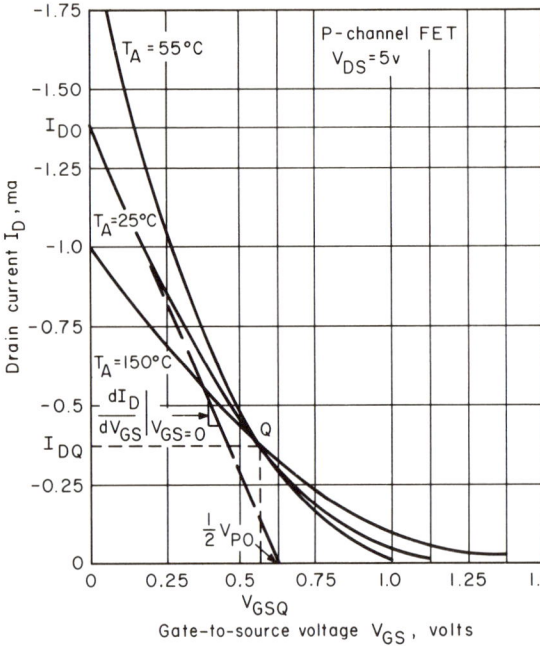

Fig. 2.4. Effect of temperature on transfer curves.

and from it we can see that a good value for n is 2. This is indeed fortunate, because it makes $M_2/n = -1$ a good approximation. Then at $T = 300°K$, dI_{DSS}/dT is zero for an FET having a pinch-off voltage of 0.6 volt.

Generally, the pinch-off voltage is difficult to obtain by direct measurements. Consider the transfer curves in Fig. 2.4. On the scale shown it is difficult to determine just where V_P is. When the channel is totally pinched-off a small amount of leakage current flows, the measurer has to observe a slowly changing current of a very small value (sometimes in the tenths of nanoamperes), and decide at what gate-to-source voltage the current has stopped changing. This is so much trouble that it is desirable to "infer" V_P by measuring I_{DSS} and g_{max} (measured at the same bias point) and using a good approximate value for M_2, e.g., $M_2 = -2$ for double-diffused FET's. V_P can also be inferred for square-law devices by measuring V_{GS} at some drain current less than I_{DSS} and calculating the value from Eq. (1.54); a convenient current for measurement is $I_{DP} = 0.1\ I_{DSS}$, and the calculated pinch-off voltage is

$$V_P = 1.46\ V_{GS}\big|_{I_{DP}=0.1 I_{DSS}}$$

Temperature Dependence of the Pinch-off Drain Current. In the preceding section, we saw that the opposite effects of the temperature of carrier mobility and junction contact potential produced a zero in the temperature coefficient of I_{DSS} on FET's with a V_P of approximately 0.6 volt. The analysis can be generalized to any pinch-off drain current by differentiating Eq. (1.14) with respect to temperature in the same manner as in Eq. (2.1):

FET Characteristics

$$\frac{dI_{DP}}{dT} = \frac{2qW}{\epsilon L} \frac{d\mu}{dT} \int_{H_S}^{\frac{1}{2}H_c} \left[Q\left(\frac{1}{2}H_c\right) - Q(H_S) \right] hp(h)\, dh$$

$$-\frac{2qW\mu}{\epsilon L}\left[Q\left(\frac{1}{2}H_c\right) - Q(H_S) \right] H_S p(H_S) \frac{dH_S}{dT} \qquad (2.13)$$

The simplification of this expression parallels that of Eq. (2.1). Again, the first term can be simplified by inspection; the simplification of the second term begins with the realization that, since V_{GS} is an external voltage, it can be assumed held constant while dI_{DP}/dT is investigated. Now

$$V_{GS} = \frac{q}{\epsilon}\int_{H_\phi}^{H_S} hp(h)\, dh = \frac{q}{\epsilon}\int_0^{H_S} hp(h)\, dh - \frac{q}{\epsilon}\int_0^{H_\phi} hp(h)\, dh \qquad (2.14)$$

This simply means that

$$V_{GS} = \frac{q}{\epsilon}\int_0^{H_S} hp(h)\, dh - \phi \qquad (2.15)$$

and

$$\frac{dV_{GS}}{dT} = \frac{q}{\epsilon} H_S p(H_S) \frac{dH_S}{dT} - \frac{d\phi}{dT} \qquad (2.16)$$

Now since we can constrain dV_{GS}/dT to zero, then

$$\frac{q}{\epsilon} H_S p(H_S) \frac{dH_S}{dT} = \frac{d\phi}{dT} \qquad (2.17)$$

Therefore Eq. (2.13) reduces to

$$\frac{dI_{DP}}{dT} = I_{DP}\frac{1}{\mu}\frac{d\mu}{dT} - g_m \frac{d\phi}{dT} \qquad (2.18)$$

Equation (2.18) can be written in terms of the FET static quantities I_{DSS} and V_P by making use of the square-law approximation of Eq. (1.54). Differentiating Eq. (1.54) with respect to V_{GS}:

$$g_m = \frac{dI_{DP}}{dV_{GS}} = 2\frac{I_{DSS}}{V_P}\left(\frac{V_{GS}}{V_P} - 1\right) \qquad (2.19)$$

$$\frac{g_m}{I_{DP}} = -\frac{2}{V_P}\sqrt{\frac{I_{DSS}}{I_{DP}}} \qquad (2.20)$$

Substituting Eqs. (2.20) and (2.8) into Eq. (2.18):

$$\frac{1}{I_{DP}}\frac{dI_{DP}}{dT} = \frac{2}{V_P}\sqrt{\frac{I_{DSS}}{I_{DP}}}\frac{d\phi}{dT} - \frac{n}{T} \qquad (2.21)$$

Equation (2.21) is the temperature coefficient of the drain current for any bias point. The zero of the temperature coefficient occurs when Eq. (2.21) equals zero, at some bias current I_{DQ}, the temperature coefficient of I_{DP} is zero when

$$I_{DQ} = I_{DSS}\frac{4T^2}{n^2V_P^2}\left(\frac{d\phi}{dT}\right)^2 \qquad (2.22)$$

Ordinarily one would expect this bias point to change with temperature, but if we assume that the free-carrier density is constant in the active channel (the square-law approximation implies this), I_{DSS} can be expressed in terms of mobility and V_P:

$$I_{DSS} = \frac{2qW}{\epsilon L}\mu V_P^2 = \frac{2qW}{\epsilon L}\mu_o\left(\frac{T}{T_o}\right)^{-n}V_{Po}^2\left[1 - \frac{1}{\phi}\frac{d\phi}{dT}(T-T_o)\right]^2 \qquad (2.23)$$

Values μ_o, T_o, and V_{Po} are reference temperature values of these quantities. Therefore, I_{DSS} can be expressed in terms of its reference temperature value I_{DO} by

$$I_{DSS} = I_{DO}\left(\frac{T}{T_o}\right)^{-n}\left[1 - \frac{1}{\phi}\frac{d\phi}{dT}(T-T_o)\right]^2 \qquad (2.24)$$

Substituting Eq. (2.24) into Eq. (2.22), we find that the bias point for zero temperature coefficient of I_{DP} is relatively independent of temperature:

$$I_{DQ} = I_{DO}\frac{4T_o^n T^{(2-n)}}{n^2 V_{Po}^2}\left(\frac{d\phi}{dT}\right)^2 \qquad (2.25)$$

Since n is approximately equal to 2, the absolute temperature practically drops out of the equation. Figure 2.4 contains a set of transfer curves of a particular FET at three different temperatures, showing that the crossover current I_{DQ} is practically independent of temperature.

Equation (2.25) is plotted in Fig. 2.5 and compared to experimental data from some 24 devices, proving that the analysis is reasonably valid.

The existence of the zero point in the temperature coefficient suggests that FET's should make very good d-c amplifiers. A more useful form of Eq. (2.21) is one that expresses the FET temperature behavior in terms of dV_{GS}/dT for a constant I_{DP}. This fits in with the widely used practice of stating the input voltage change of a d-c amplifier necessary to maintain a constant output. The derivation of dV_{GS}/dT is done in the same manner as the derivation of Eq. (2.21) except that I_{DP} and not V_{GS} is held constant by external means; i.e., the FET is imagined to be supplied from a constant current source. The algebra is left to the interested reader; the result is

$$\frac{dV_{GS}}{dT} = \frac{dV_P}{dT} - \frac{nV_P}{2T}\sqrt{\frac{I_{DP}}{I_{DSS}}} \qquad (2.26)$$

This equation is plotted in Fig. 2.6: dV_{GS}/dT is plotted against I_{DP}/I_{DSS}, with V_P as the

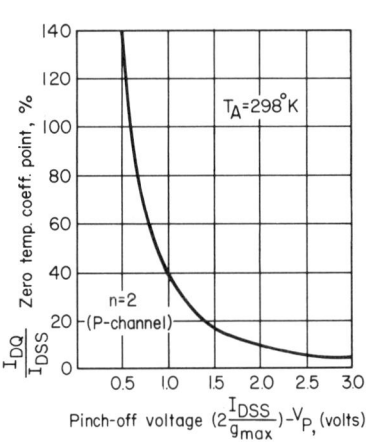

Figure 2.5

FET Characteristics

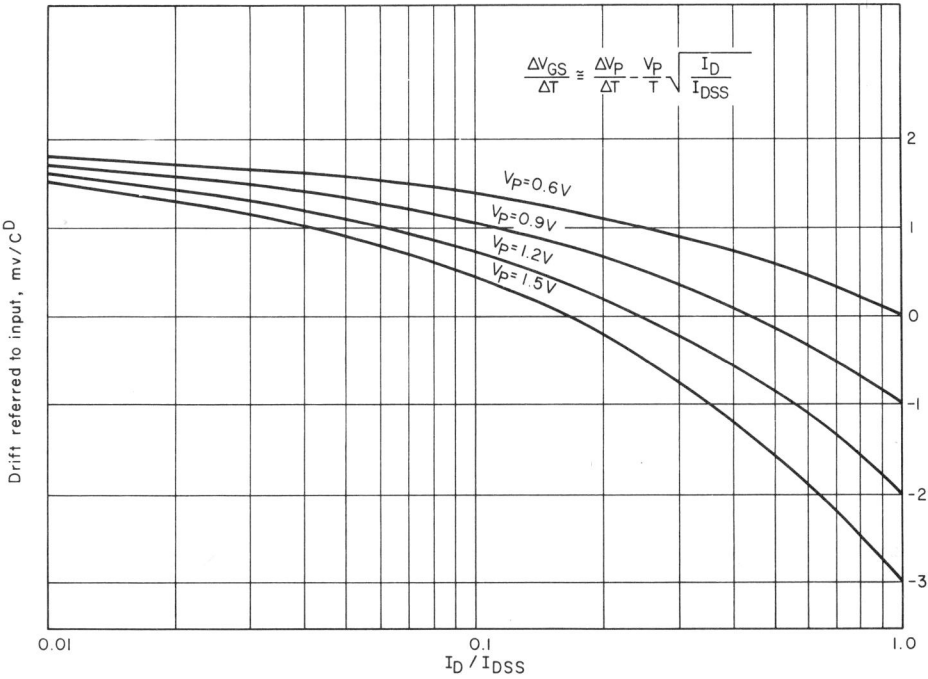

Figure 2.6

running parameter. The zero crossings are those described by Eq. (2.26); for example, $dV_{GS}/dT = 0$ at $V_P = 0.6$ volt, and $I_{DP}/I_{DSS} = 1$. At very low drain currents, the temperature coefficient of V_{GS} approaches dV_P/dT, about 2 mv/C°.

Breakdown Voltage. In Chap. 1 we saw that the breakdown mechanism in FET's is avalanche breakdown of the gate-to-channel diode. We also saw that reverse voltage on this diode varies along the gate length, being highest at the drain end of the channel. In an FET with no manufacturing flaws, this is where the breakdown occurs. If the drain and source terminals are interchanged, the breakdown voltage should be about the same. Since the reverse bias between the gate and source must be increased to reduce drain currents, it follows that, for a fixed drain-to-source supply voltage, the apparent drain breakdown voltage must decrease monotonically with current. Take the characteristics in Fig. 2.1; at $V_{GS} = 0$ the breakdown voltage occurs at $V_{DS} = -27$ volts (P-channel device). Now remember that the breakdown actually occurs between the gate and drain terminals; i.e., the high current flows in the gate terminal, and since the gate is at zero volts (with respect to the source) the drain-to-gate breakdown is -27 volts. Now if V_{GS} is increased to $+0.2$ volt, the drain-to-gate breakdown voltage is still -27 volts, but the apparent drain-to-source breakdown voltage is now -26.8 volts. The limiting breakdown voltage of the device is the drain-to-gate breakdown voltage. This theoretically will be the same regardless of the source potential which can vary between the source being tied to the drain at one extreme, and being tied to the gate at the other. This voltage is usually specified with the source open, and is called BV_{DGO} according to convention. When the source is operated at a given bias voltage of X volts, the apparent drain-to-source breakdown voltage is BV_{DSX}, and is related to BV_{DGO} by

$$BV_{DSX} = BV_{DGO} + V_{GSX} \qquad (2.27)$$

Leakage Currents. The gate and channel in a unipolar FET form a PN junction; the diode currents and voltages are governed by the usual exponential law:

$$I_G = I_o \left[\epsilon \frac{qV_{GC}}{KT} - 1 \right] \qquad (2.28)$$

where I_G = gate current
I_o = reverse saturation gate current
V_{GC} = gate-to-channel voltage, positive in forward bias
K = Boltzmann's constant, joules/K°
T = absolute temperature, °K

When the gate is used as the input terminal of an FET, the dynamic input conductance is determined by the slope of the input-current voltage characteristic shown in Fig. 2.7, which is a qualitative plot of Eq. (2.28) in the vicinity of the origin. dI_G/dV_{GC} from Eq. (1.26) is

$$\frac{dI_G}{dV_{GC}} = \frac{q(I_G - I_o)}{KT} \qquad (2.29)$$

KT/q at room temperature is about 25 m, while the I_o of a unipolar FET might be as low as 10^{-10} amps. At $I_G = 0$ ($V_{GC} = 0$), the input resistance dV_{GC}/dI_G at room temperature of such an FET will be as high as 250 megohms, and will increase with the reverse bias. Since the FET is normally operated with a reverse-biased gate, it is obviously a very-high-input-resistance device. The equivalent circuit of the input terminals of a normally biased FET is a very-high-value nonlinear resistor shunted by a capacitor (in the range of 1 to 50 pf for present devices), and a very small constant-current generator. Equation (2.29) indicates that the nonlinear input resistance can become very small if the gate is allowed to become forward-biased, a problem to which the surface-type FET's are not subject.

Channel ON Resistance, $r_{D(on)}$. The expression $r_{D(on)}$, the value of the drain-to-source resistance when $V_{GS} = 0$, is the slope of the zero bias output curve (V_{DS} vs. I_D) at the origin. It is by definition equal to $1/g_{max}$; remember that g_{max} is the conductance of a parallelepiped section of the channel bounded by the depletion layers at the source end. When $r_{D(on)}$ is measured, there is no average current flow and the channel height modulation effects are negligible. The generalized

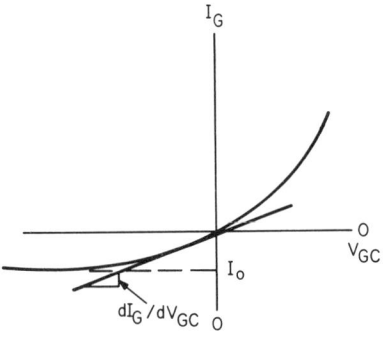

Figure 2.7

zero drain voltage channel resistance at any bias is $r_D = 1/g_m$; therefore, from Chap. 1:

$$r_D = \frac{V_P}{2I_{DSS}\left(\dfrac{V_{GS}}{V_P} - 1\right)} \tag{2.30}$$

This resistance can be reduced to a value less than $r_{D(on)}$ simply by forward-biasing the gate. The minimum possible value of r_D is

$$r_{D(min)} = r_{D(on)} \frac{V_P}{V_P + \phi} \tag{2.31}$$

2.2 DYNAMIC CHARACTERISTICS

Small-signal Low-frequency Behavior. FET channel current actually flows in a nonlinear distributed RC transmission line, but for low-frequency work it is sufficient to treat the device as if it were a lumped nonlinear electrical network. An equivalent circuit showing the "lumps" necessary for low-frequency calculations is given in Fig. 2.8. Capacitances C_{DG} and C_{SG} and conductances g_{DG} and g_{SG} are a lumped-element representation of the reverse-biased gate-to-channel diode. In a well-designed FET, g_{DG} and g_{SG} will be quite small and can be considered open circuits. Values r_{DB} and r_{SB} are the bulk resistances of the semiconductor path from the channel edges to the drain and source contacts respectively; they will be in the order of 100 ohms or less, depending somewhat on geometry and manufacturing process. At low frequencies, the effect of r_{DB} is quite negligible; it can be considered as only a very small part of any practical load resistance. The value of r_{SB} has a slight, generally negligible effect on the apparent transconductance of the device; the voltage $v_{gs'}$ in Fig. 2.8 is related to the terminal voltage v_{gs} by

$$v_{gs'} = \frac{v_{gs}}{1 + g_m r_{SB}} \tag{2.32}$$

For an FET having a g_m of 1,000 μmhos and an r_{SB} of 75 ohms, the denominator of Eq. (2.32) is 1.075. The value g_{DS} is the slope of the output characteristics in the pinch-off region and is generally small compared to practical load conductances.

It is common practice among manufacturers to specify the FET by its equivalent short-circuit admittance parameters. The general two-port y parameter network

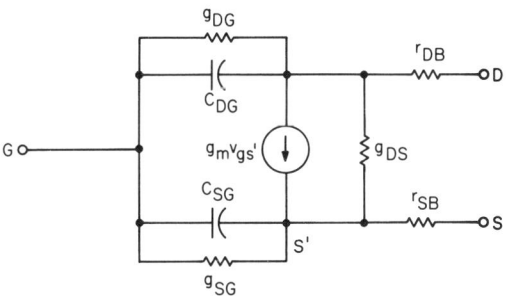

Figure 2.8

is shown in Fig. 2.9. The terminal small-signal voltages and currents are referred to the common-source connection. The parametric equations of the network are:

$$i_g = y_{is}v_{gs} + y_{rs}v_{ds} \tag{2.33}$$

$$i_d = y_{fs}v_{gs} + y_{os}v_{ds} \tag{2.34}$$

The terminal conditions for determining the parameters are:

$$\text{Output shorted:} \quad y_{is} = \frac{i_g}{v_{gs}}$$

$$y_{fs} = \frac{i_d}{v_{gs}}$$

$$\text{Input shorted:} \quad y_{rs} = \frac{i_g}{v_{ds}}$$

$$y_{os} = \frac{i_d}{v_{ds}}$$

When these conditions are applied to the physical equivalent circuit in Fig. 2.8, the y parameters can be written in terms of the lumped physical elements. If all the diode conductances and bulk resistances of Fig. 2.8 are neglected, the y parameters are:

$$y_{is} = j\omega(C_{DG} + C_{SG}) \tag{2.35}$$

$$y_{rs} = -j\omega C_{DG} \tag{2.36}$$

$$y_{fs} = g_m - j\omega C_{DG} \tag{2.37}$$

$$y_{os} = g_{DS} + j\omega C_{DG} \tag{2.38}$$

These parameters are all bias dependent. We have seen that the bias dependence of g_m in the pinch-off region is easily obtained by differentiating the square-law approximation of I_{DP} with respect to v_{GS}:

$$\frac{dI_{DP}}{dv_{GS}} = g_m = \frac{2I_{DSS}}{V_P}\left(\frac{v_{GS}}{V_P} - 1\right) \tag{2.39}$$

If it is desirable to express g_m in terms of I_{DP} rather than v_{GS}, some simple algebra yields

$$g_m = \frac{2}{V_P}\sqrt{I_{DSS}I_{DP}} \tag{2.40}$$

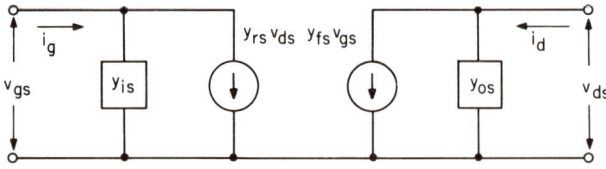

Figure 2.9

We discovered in Chap. 1 that obtaining an analytical expression for the bias dependence of g_{DS} is a hopeless task; our only choice is to rely on point-to-point measurements supplied by the manufacturer or to assume that the load conductance is large and neglect g_{DS}.

If it can be assumed that the total gate-to-channel capacitance C_G, discussed in Chap. 1, is equal to the sum $(C_{DG} + C_{SS})$, then this sum can be expressed in terms of FET constants and bias-dependent terminal characteristics. In Eq. (1.44) we saw that

$$C_G = 2\epsilon W \int_0^L \frac{dx}{h(x)}$$

and from Eq. (1.47):

$$dx = \frac{2qW\mu}{I_{DP}\epsilon}\left[Q\left(\frac{1}{2}H_c\right) - Q(h)\right]hp(h)\,dh$$

Substituting Eq. (1.47) into Eq. (1.44) we find that

$$C_G = 2\epsilon W \frac{2qW\mu}{\epsilon I_{DP}} \int_{H_S}^{\frac{1}{2}H_c} \left[Q\left(\frac{1}{2}H_c\right) - Q(h)\right]p(h)\,dh \tag{2.41}$$

Now from Eq. (1.6):

$$p(h)\,dh = \frac{1}{q}dQ(h) \tag{2.42}$$

Substituting this into Eq. (2.41) we have

$$C_G = \frac{4W^2\mu}{I_{DP}} \int_{Q(H_S)}^{Q(\frac{1}{2}H_c)} \left[Q\left(\frac{1}{2}H_c\right) - Q(h)\right]dQ(h) \tag{2.43}$$

Equation (2.43) is integrable directly; the result of the integration is

$$C_G = \frac{2W^2\mu}{I_{DP}}\left[Q\left(\frac{1}{2}H_c\right) - Q(H_S)\right]^2 \tag{2.44}$$

which, from Eq. (1.25), can be written in terms of g_m:

$$C_G = \frac{g_m^2}{2I_{DP}}\frac{L^2}{\mu} \tag{2.45}$$

L and μ are constants. L is determined for double-diffused FET's by photomasking techniques; an approximate value for FET's of the 2N2497 and 2N3329 types is about 0.5 mil. The value of μ for holes in P-type silicon is 500 cm²/volt-sec. Equation (2.45) holds for an FET having the ideal geometry of Fig. 1.5. The double-diffused FET of Fig. 1.17 has a large-area "back" or substrate gate which will make the total gate capacitance somewhat larger than predicted. In the future, expanded-contact techniques will foreshorten the back gate considerably and bring the gate capacitance more nearly into line with calculations based on the ideal model.

One is tempted to apply the square-law approximation expressions for I_{DP} and g_m to Eq. (2.45) for further simplification. This can be done, but we must be care-

44   Field-effect Transistors

ful about interpreting the result. Remember that the square-law approximation becomes exact only for channels whose height is small compared to the height of the depletion layer caused by the contact potential. It is exact for the limiting case of a channel of infinitesimal height, our spike profile in Chap. 1. Therefore, we must expect no change in the gate-to-channel capacitance C_G with bias if the square-law approximation is applied to Eq. (2.45). To obtain bias dependence of C_G, g_m, and I_{DP} must be evaluated from the geometry and doping level profiles of Chap. 1. We apply the square-law approximation here because it is useful in providing a quick check on the frequency capabilities of FET's about which the manufacturer chooses only to specify static characteristics on his data sheet. A few such data sheets are still in circulation, but the new EIA registration format should eventually eliminate them.

Applying square-law approximations for g_m and I_{DP} to Eq. (2.45), we have

$$C_G = \frac{2I_{DSS}}{V_P^2}\frac{L^2}{\mu} \approx C_{DG} + C_{SG} \qquad (2.46)$$

High-frequency Second-order Effects. The bulk resistances r_{DB} and r_{SB} in Fig. 2.8 will add a second-order term to the zero and first-order approximations given for the FET y parameters in Eqs. (2.35) through (2.38). The simple example in Fig. 2.10 makes the reason clear. A fundamental theorem of network analysis states that any series circuit, such as the one in Fig. 2.10a, can be replaced at a single frequency by any equivalent parallel circuit like the one in Fig. 2.10b. At any frequency

$$y_P = \frac{1}{Z_S} \qquad (2.47)$$

where

$$y_P = \frac{1}{R_P} + j\omega C_P$$

and

$$Z_S = R_S - j\frac{1}{\omega C_S}$$

When the complex notation is substituted for y_P and Z_S and the real and imaginary parts are equated, we have

$$\frac{1}{R_P} = \frac{\omega^2 C_S^2 R_S}{1 + \omega^2 C_S^2 R_S^2} \qquad (2.48)$$

$$C_P = \frac{C_S}{1 + \omega^2 C_S^2 R_S^2} \qquad (2.49)$$

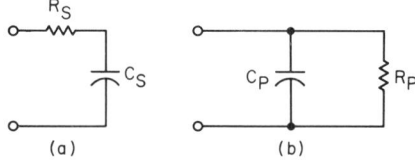

Figure 2.10

FET Characteristics

at frequencies where $\omega^2 C_S^2 R_S^2 \ll 1$, y_P can be approximated by

$$y_P = \omega^2 C_S^2 R_S + j\omega C_S \tag{2.50}$$

Thus the real part of y_P is a frequency-squared or second-order term, and the imaginary part is a first-order term dependent on the first power of the angular frequency. When the effects of r_{DB} and r_{SB} of Fig. 2.8 are taken into account in the expressions for the y parameters and suitable simplifying assumptions are made y_{is}, y_{rs}, and y_{os} can be rewritten to include the second-order effects almost by inspection:

$$y_{is} = \omega^2(C_{DG}^2 r_{DB} + C_{SG}^2 r_{SB}) + j\omega(D_{DG} + C_{SG}) \tag{2.51}$$

$$y_{rs} = -\omega^2 C_{DG}^2 r_{DB} - j\omega C_{DG} \tag{2.52}$$

$$y_{os} = g_{DS} + \omega^2 C_{DG}^2 r_{DB} + j\omega C_{DG} \tag{2.53}$$

These approximations serve very well to explain the behavior of the y parameters at high frequencies, and they do not severely complicate the model. Plots of these three y parameters are given in Figs. 2.11 to 2.13, for a 2N2499. Note that the imaginary parts slope at 20 db per decade and the real parts ultimately slope at 40 db per decade, because of the first-order and second-order frequency terms

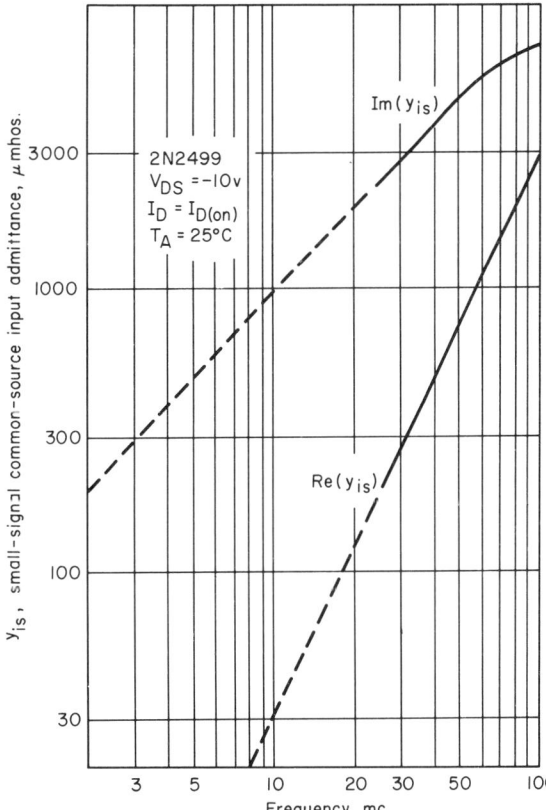

Fig. 2.11. Small-signal common-source input admittance vs. frequency.

46 Field-effect Transistors

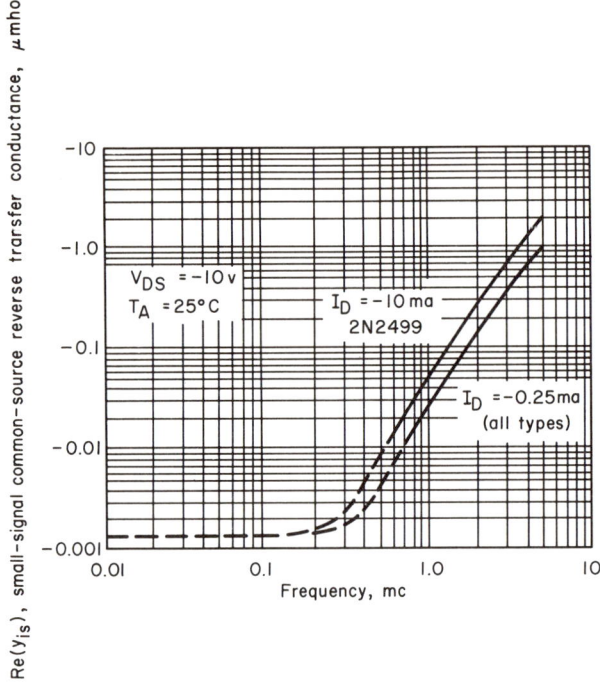

Fig. 2.12. Small-signal common-source reverse transfer conductance vs. frequency.

respectively. The numerical values on the graphs can be verified by using the 2N2499 data sheet; r_{DB} and r_{SB} are each about 50 ohms for the 2N2499.

Noise Characteristics. Three principal noise-generating mechanisms are operative in FET's:

1. The gate leakage current, which generates full shot noise.
2. A thermal noise voltage generated in the conducting channel which modulates the space-charge-layer height.
3. Generation-recombination noise in the space-charge layer, $1/f$ in character, which also modulates the height of the space-charge layer.

The shot-noise current due to random carrier collection across the PN junction is given by

$$\overline{i_{sh}^2} = 2qi_G \Delta f \qquad (2.54)$$

where i_G = forward *or* reverse gate current

q = the electronic charge, 1.6019×10^{-19} coulomb

Δf = effective noise bandwidth, cps

In unipolar FET's operated normally with a reverse-biased gate, the reverse leakage of the diode can range from 10^{-8} down to 10^{-10} amp. In surface FET's, i_G is very small indeed, 10^{-15} amps or lower, so that this noise source is not a factor and can be ignored.

FET Characteristics

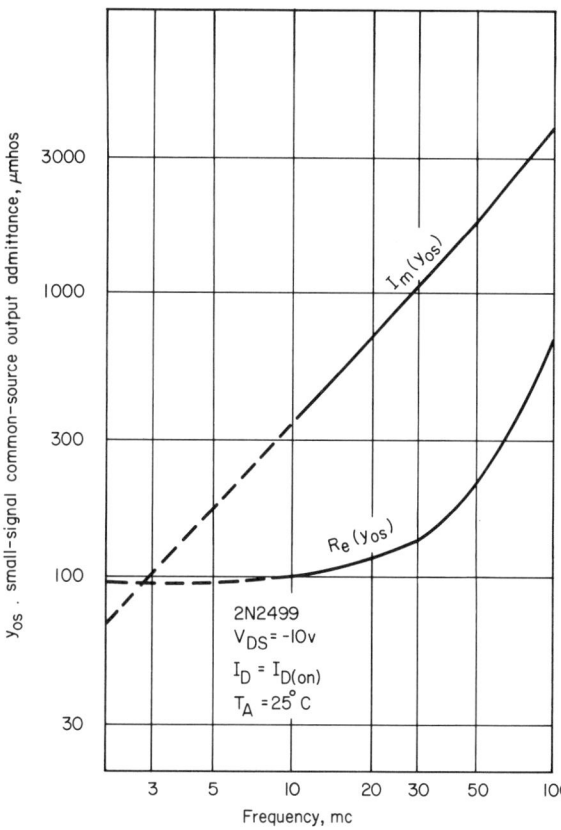

Fig. 2.13. Small-signal common-source output admittance vs. frequency.

The thermal noise generated in the conducting channel has been shown by van der Ziel[1] to be approximately the equivalent of the noise contributed by an external source resistor equal to $1/g_m$. The thermal noise voltage generated in such a resistor is

$$\overline{e_{th}^2} = 4KT \frac{1}{g_m} \Delta f \qquad (2.55)$$

The generation-recombination noise in the space-charge layer effectively modifies the thermal noise voltage in Eq. (2.55) by giving it a $1/f$ frequency dependence:

$$\overline{e_f^2} = \overline{e_{th}^2} \left(1 + \frac{f_{c1}}{f}\right) \qquad (2.56)$$

where f_{c1} = corner frequency, or turnover frequency at the low end of the spectrum.

These physical mechanisms suit the FET to characterization by the so-called two-generator method. The theorem that forms the basis for this method of characterization is that any linear active two-port network can be represented by a series noise-voltage generator and a shunt noise-current generator at the input of an ideal noiseless network.

48 Field-effect Transistors

In Fig. 2.14, the example illustrated is a linear amplifier. The method of characterization is good only for small signals, but since noise signals are generally small signals, there is no problem. The term γ is the correlation coefficient between the two generators, and indicates the degree of independence or dependence of the two generators. In other words, it tells how much one generator is affected by variations in the other; if γ is one, there is a functional relationship between the two generators; if γ is zero, the two generators are completely independent of each other.

Fortunately for FET's, there is little correlation between the shot noise in the gate leakage current and the thermal noise in the channel.

Measurement of e_n and i_n is straightforward, but that of i_n is quite difficult. For measurements of e_n, the gate terminal of the FET is returned to the source through a small resistance, preferably an ideal short, that satisfies the following conditions:

$$R_{short} \ll R_{in}$$
$$i_n R_{short} \ll e_n$$

The first condition assures that all of the generator voltage e_n will appear across the amplifier input, while the second condition limits the amount of signal current contributed by i_n. The output current of the FET contains a noise component which is the effective thermal noise voltage generated in the channel and the space-charge layer times g_m. Then to obtain e_n, we simply divide by g_m:

$$e_n = \frac{i_{dn}}{g_m} = \sqrt{\overline{e_{th}^2}\left(1 + \frac{f_{cl}}{f}\right)} \tag{2.57}$$

Data available on the 2N2500 (32 kc for f_{cl}) makes Eq. (2.57) a reasonably accurate description of the equivalent input noise voltage of the 2N2500, and, in

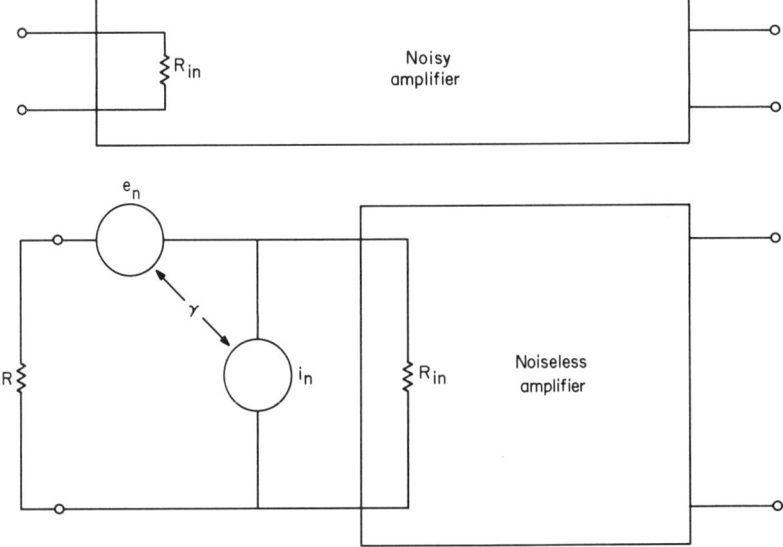

Figure 2.14

FET Characteristics

general, of most double-diffused FET's. This f_{cl} is not be be confused with the corner frequency of the *noise figure*.

In the measurement of i_n, it is necessary to open-circuit the input terminals of the FET to such a degree that

$$R_{open} \gg R_{in}$$
$$i_n R_{open} \gg e_n$$

The first requirement is extremely difficult to satisfy; in most cases, it is sufficient to measure the gate leakage current at the desired bias condition and calculate i_n using Eq. (2.54), where $i_n \approx (i_{sh}^2)^{1/2}$ and $i_G = I_{GSX}$, the biased static gate current.

Noise performance of the FET in any circuit can be calculated if we know the values of the two generators, simply by substituting an equivalent circuit for the FET into the "Noiseless amplifier" block in Fig. 2.14. The two generators can also be used to obtain the more familiar but less useful, noise factor, and hence noise figure, of the FET in a given circuit. The noise factor of a network depends on the resistance of the signal generator that drives the network and is defined as

$$F = \frac{\text{available noise power output}}{\text{available power output due to thermal noise in } R_g}$$

where R_G is the internal resistance of the signal generator. By this definition, the noise factor is

$$F = 1 + \frac{1}{4KT\Delta f}\left(\overline{i_n^2} + \frac{\overline{e_n^2}}{R_G}\right) \qquad (2.58)$$

In terms of FET parameters, the noise factor is

$$F = 1 + \frac{qI_{GSX}R_g}{2KT} + \frac{1 + \dfrac{f_{cl}}{f}}{g_m R_g} \qquad (2.59)$$

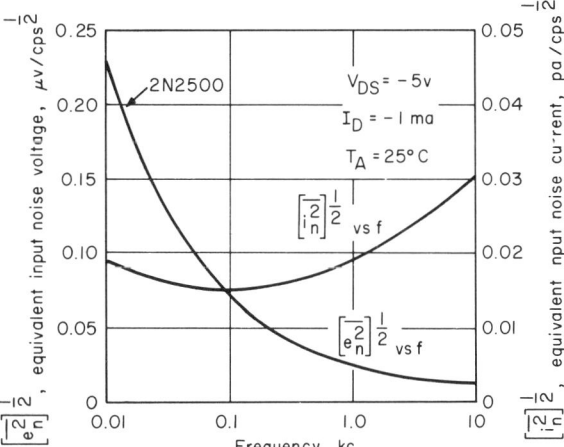

Fig. 2.15. Equivalent input noise voltage and equivalent input noise current vs. frequency.

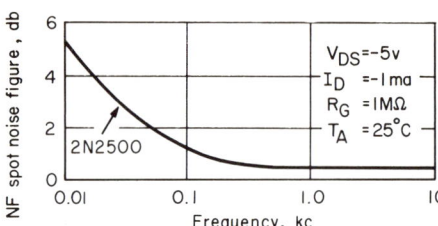

Fig. 2.16. Spot noise figure vs. frequency.

Equation (2.59) demonstrates why FET noise figures (noise figure = $10 \log_{10} f$) have such low corner frequencies especially for high values of R_g. Figure 2.15 shows typical values of e_n and i_n vs. frequency for a 2N2500 biased at 1-ma drain current. The high corner frequency of e_n is apparent; contrast this with the graph in Fig. 2.16, which is a plot of the noise figure vs. frequency for the 2N2500 at the same bias, with a 1-megohm signal generator resistance.

BIBLIOGRAPHY

1. Shockley, W.: "Electrons and Holes in Semiconductors," D. Van Nostrand Company, Inc., Princeton, N.J., 1950.
2. van der Ziel, A.: Thermal Noise in Field-effect Transistors, *Proc. IRE,* vol. 50, pp. 1808–1812, August, 1962.

3

FET's in Low-level Linear Circuits

3.1 BASIC FET AMPLIFIER CIRCUITS

The FET, like the other three-terminal active networks, can be operated with any of its three terminals common in an amplifier circuit. The three basic amplifier configurations are shown in Fig. 3-1a–c which are common-source, common-gate, and common-drain amplifiers respectively. In each of these connections it is possible to reverse the input and output terminal pairs, giving altogether six possible combinations, but the three connections shown are the only ones that have any major applications.

The FET symbol shown in Fig. 3.1 indicates the interchangeability of the drain and the source terminals. Without some special marking, the drain is indistinguishable from the source. Hereafter, we will designate the drain terminal by a D on the symbol as in Fig. 3.1. The symbol shown indicates a P-channel FET; for an N-channel FET, the arrow on the gate terminal is reversed.

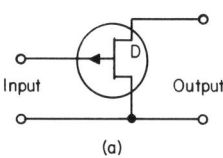

(a)

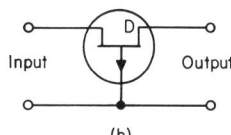

(b)

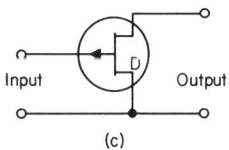

(c)

Fig. 3.1. Basic amplifier configurations: (a) common source, (b) common gate, (c) common drain.

The Common-source Amplifier. The circuit in Fig. 3.2 is a basic common-source FET amplifier with two fixed-bias batteries. A signal voltage generator v_g is applied to the input and the output is taken between the drain terminal and the common terminal. Note that the polarities of V_{GG} and V_{DD} are just the opposite of what we are accustomed to in vacuum-tube circuits; the P-channel FET has in fact been facetiously called a "PNP vacuum tube." For an N-channel FET, of course, the polarities are reversed. Besides the waste of an extra power supply, using fixed bias as shown in Fig. 3.2 is not generally desirable because FET characteristics vary so much with temperature and device selection. It may be possible to adjust V_{GG} to set the drain current on an individual FET to the point of zero drift, derived in Chap. 2; but chances are good that V_{GG} would have to be readjusted every time the device is changed in

51

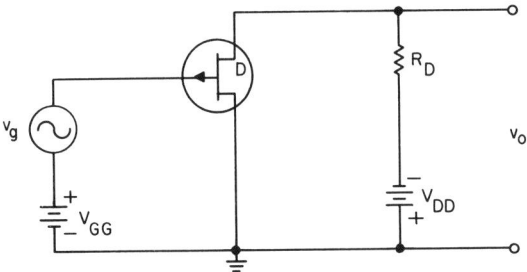

Fig. 3.2. Basic common-source FET amplifier.

a given amplifier circuit. For these reasons, in most practical circuits, some form of bias is used other than a fixed V_{GG} between the gate and the source. The circuit in Fig. 3.3 shows a method of obtaining a reverse voltage between the gate and source by using the voltage drop across a resistor in the source lead; this voltage drop is caused by the FET's own source current. Such a scheme is called *self-bias*. This is still a common-source amplifier at signal frequencies above which the reactance of the source bypass capacitor C_S is negligible compared to the resistance of R_S. More will be said about biasing problems later.

Circuit calculations can be made on a common-source amplifier at signal frequencies of interest by replacing the FET in Fig. 3.2 by the low-frequency equivalent circuit of Fig. 2.9, and assuming that C_S and the supply voltage V_{DD} are short circuits to alternating current. The equivalent circuit of the amplifier is given in Fig. 3.4; for the moment we will assume operation at frequencies low enough so that the effects of C_{DG} and C_{SG} are negligible. Z_L includes the drain-bias resistor R_D and any other a-c load on the output terminals, such as the input impedance of another amplifier. The output voltage of the amplifier is

$$v_o = -g_m v_{gs} \frac{Z_L}{1 + Z_L g_{DS}} \tag{3.1}$$

Therefore the voltage gain of the amplifier is

$$A = \frac{v_o}{v_{gs}} = -\frac{g_m Z_L}{1 + Z_L g_{DS}} \tag{3.2}$$

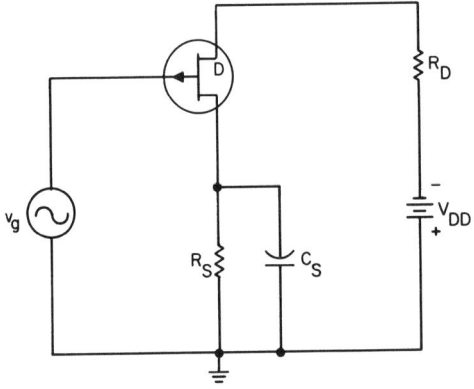

Fig. 3.3. Self-biased amplifier.

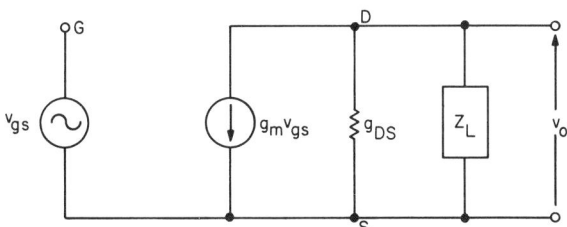

Fig. 3.4. Low-frequency equivalent circuit.

and when $Z_L g_{DS} \ll 1$, the gain reduces to

$$A = -g_m Z_L \tag{3.3}$$

which is quite familiar to vacuum-tube engineers as the approximate gain of a pentode stage. The output admittance of this amplifier stage at low frequencies is g_{DS}, again just like the pentode vacuum tube.

The effect of leaving part of the source biasing resistance unbypassed on the voltage gain can be determined from the circuit in Fig. 3.5. In this circuit, the load current i_L is

$$i_L = -g_m v_{gs} \frac{1}{1 + g_{DS}(Z_L + R_S)} \tag{3.4}$$

Now

$$v_{gs} = v_g + i_L R_S \tag{3.5}$$

then

$$i_L = -\frac{g_m v_g}{1 + g_m R_S + g_{DS}(Z_L + R_S)} \tag{3.6}$$

and

$$v_o = i_L Z_L = -\frac{g_m v_g Z_L}{1 + g_m R_S + g_{DS}(Z_L + R_S)} \tag{3.7}$$

The modified voltage amplification is now

$$A = -\frac{g_m Z_L}{1 + g_m R_S + g_{DS}(Z_L + R_S)} \approx -\frac{g_m Z_L}{1 + g_m R_S} \tag{3.8}$$

Comparison of Eq. (3.8) with Eq. (3.3), shows that A is reduced by approximately $1/(1 + g_m R_S)$ when part of the source resistance is left unbypassed.

Input Admittance of a Common-source Amplifier. If the capacitances are added to the low-frequency equivalent circuit, the loading that the FET presents to

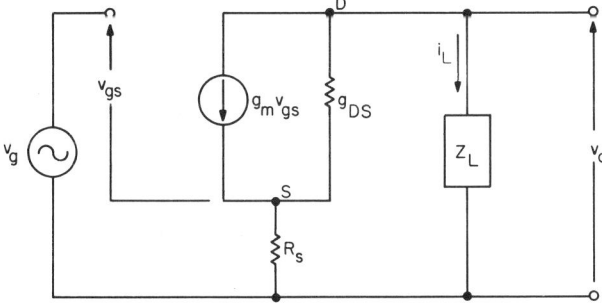

Fig. 3.5. Amplifier circuit with a source resistor.

54 Field-effect Transistors

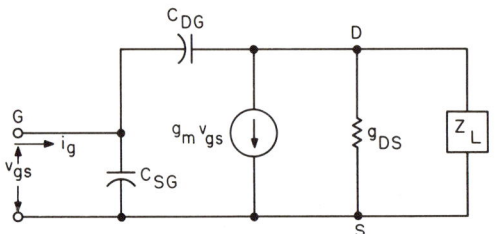

Fig. 3.6. Common-source equivalent circuit with capacitances.

a generator can be calculated, at least at frequencies where bulk resistance second-order effects are negligible. The gate signal current i_g in Fig. 3.6 is

$$i_g = j\omega C_{SG} v_{gs} + j\omega C_{DG}(v_{gs} - v_o) \qquad (3.9)$$

Substitution of Eq. (3.1) for v_o and some simple manipulation yield the input admittance of a common-source FET amplifier:

$$y_{in} = \frac{i_g}{v_{gs}} j\omega \left[C_{SG} + C_{DG} 1 + g_m \left(\frac{Z_L}{1 + g_{DS} Z_L} \right) \right] \qquad (3.10)$$

If Z_L is purely resistive, the input admittance looks capacitive, having a value

$$C_{in} = C_{SG} + C_{DG}(1 - A) \qquad (3.11)$$

The drain-to-gate capacitance appears multiplied by one plus the voltage gain. Thus we see that the FET in common source is subject to the Miller effect just as the vacuum tube is in common-cathode circuits, and as the transistor is in common-emitter circuits.

If Z_L has a reactive component, it is obvious from Eq. (3.10) that y_{in} will contain a real part. If the reactance is capacitive, the real part of y_{in} is always positive, and if the reactance is inductive, the real part of y_{in} is negative. In this case, the amplifier is unstable and will oscillate at the frequency at which the imaginary part of y_{in} is zero.

Since most data sheets give the short-circuit y parameters, Eq. (3.10) will be in more convenient form when expressed in terms of the y parameters, thus

$$y_{in} = j\omega(C_{is} - AC_{rs}) \qquad (3.12)$$

Common-gate Amplifier. Although this configuration seldom finds application, it will be interesting to go through the gain and impedance calculations. The equivalent circuit is connected for common-gate operation in Fig. 3.7. The voltage gain can be calculated by summing currents at the output node:

$$g_m v_{SG} = -v_{sg} g_{DS} + v_o \left(g_{DS} + \frac{1}{Z_L} + j\omega C_{DG} \right) \qquad (3.13)$$

The voltage gain is

$$A = \frac{v_o}{v_{sg}} = \frac{(g_m + g_{DS}) Z_L}{1 + Z_L(g_{DS} + j\omega C_{DG})} \qquad (3.14)$$

At low frequencies, the magnitude of the common-gate gain is approximately equal to that of the common-source gain (provided that $g_{DS} \ll g_m$) but unlike the

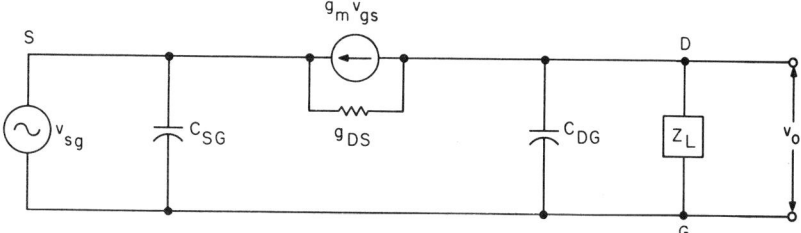

Fig. 3.7. Common-gate equivalent circuit.

common-source gain, there is no sign reversal. For sinusoidal signals, a sign reversal means 180 degrees of phase shift. Thus, in a common-gate amplifier, the the output is in phase with the input (at frequencies where reactances are negligible) and in a common-source amplifier, the output is 180 degrees out of phase with the input.

The input admittance of a common-gate amplifier can be determined by summing currents at the input node of Fig. 3.7.

$$i_g = j\omega C_{sg}v_{sg} + g_m v_{sg} + j\omega C_{DG}(v_{sg} - v_o) \quad (3.15)$$

therefore,

$$y_{in} = \frac{i_g}{v_{sg}} = g_m + g_{DS}(1 - A) + j\omega C_{SG} \quad (3.16)$$

In a common-gate stage, y_{in} has a large real part, the input resistance is approximately $1/g_m$, and it does not exhibit Miller effect. We see now why this configuration finds little application: there is no need for it. If low input impedance is what we seek, we have transistors. Nor does the lack of Miller effect offer any real advantage: many high-frequency transistors will give good power gain at frequencies at least an order of magnitude higher than will the common-gate FET.

Common-drain Amplifier or Source-Follower. This connection is analogous to the vacuum-tube cathode follower and the transistor emitter-follower, and is very useful for impedance transformations when FET's are being used with bipolar transistors.

The voltage gain of a source-follower can be determined from the equivalent circuit in Fig. 3.8. The necessary equations are obtained by summing currents at the source node and summing voltages around the input loop:

$$g_m v_{gs} = v_o \left(g_{DS} + \frac{1}{R_S} + j\omega C_{SG} \right) - j\omega C_{SG} v_g \quad (3.17)$$

$$v_{gs} = v_g - v_o \quad (3.18)$$

Substituting Eq (3.18) into Eq. (3.17) and rearranging:

$$A = \frac{v_o}{v_g} = \frac{(g_m + j\omega C_{SG})R_S}{1 + (g_m + g_{DS} + j\omega C_{SG})R_S} \quad (3.19)$$

At low frequencies, the gain reduces to approximately

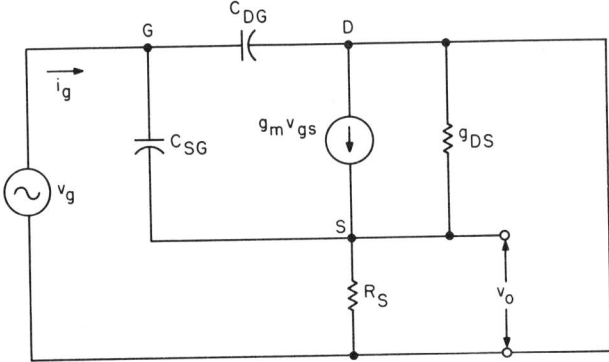

Fig. 3.8. Source-follower equivalent circuit.

$$A \approx \frac{g_m R_S}{1 + (g_m + g_{DS})R_S} \quad (3.20)$$

The maximum possible source-follower gain is

$$A_{max} = \frac{g_m}{g_m + g_{DS}} \quad (3.21)$$

for a very large R_S. We should recognize this equation from vacuum-tube analysis as being $\mu/(\mu + 1)$, where μ is the voltage amplification factor and is not to be confused with mobility. The gain is then nearly unity and there is no phase shift; the source "follows" the gate.

The input impedance of a source-follower can be obtained by first summing currents at the gate node:

$$i_g = j\omega v_g(C_{DG} + C_{SG}) - j\omega C_{SG} v_o \quad (3.22)$$

Dividing through by v_g,

$$y_{in} = \frac{i_g}{v_g} = j\omega[C_{DG} + C_{SG}(1 - A)] \quad (3.23)$$

The effect of C_{SG} is drastically reduced, especially if $A \approx 1$, but C_{DG} appears in shunt undiminished. There is no real part other than the diode shunt conductances g_{DG} and g_{SG} that we previously neglected, and the high-frequency second-order effects. The source-follower has a very high input resistance shunted by a capacitor whose capacitance is considerably less than the full capacitance of the gate-to-channel diode. Later we shall see a method for reducing C_{DG} to obtain an amplifier with a very low input capacitance.

Of particular interest in a source-follower is its output impedance, which we can calculate by setting $v_g = 0$ in Fig. 3.8 and summing currents at the source node. This procedure yields

$$i_o = v_o \left(\frac{1}{R_S} + g_{DS} + j\omega C_{SG}\right) + g_m v_o \quad (3.24)$$

so that

$$Z_o = \frac{v_o}{i_o} = \frac{R_S}{1 + (g_{DS} + g_m)R_S + j\omega C_{SG}} \quad (3.25)$$

For large values of R_S, the output resistance is approximately

$$R_o \approx \frac{1}{g_m} \quad (3.26)$$

3.2 BIASING FET AMPLIFIERS

Setting the bias of a vacuum-tube amplifier, in most cases, is simply a matter of obtaining a set of output characteristics, selecting a desired operating point such as point Q in Fig. 3.9, and drawing a load line from point Q to the supply voltage on the abscissa. The slope of this line is equal to the reciprocal of the necessary plate load resistor. The grid voltage line at Q determines the grid bias supply for fixed bias, or if self-bias is used, the grid voltage divided by the plate quiescent current I_{PQ} yields the size of the cathode resistor.

When this approach is used to bias a transistor amplifier, the results are often catastrophic. Since transistor parameters vary so much, a more stable bias circuit is obtained when the design is done analytically, taking into account "worst-case" transistor parameters.

A reasonable approach to FET biasing borrows features from both vacuum-tube and transistor biasing. Because of the presence of a well-behaved transfer charactersitic, a simple approach to bias design can be worked out using graphic and analytical methods. We will not go into the details of a worst-case bias design because such a procedure is tedious and boring and can be worked out by any col-

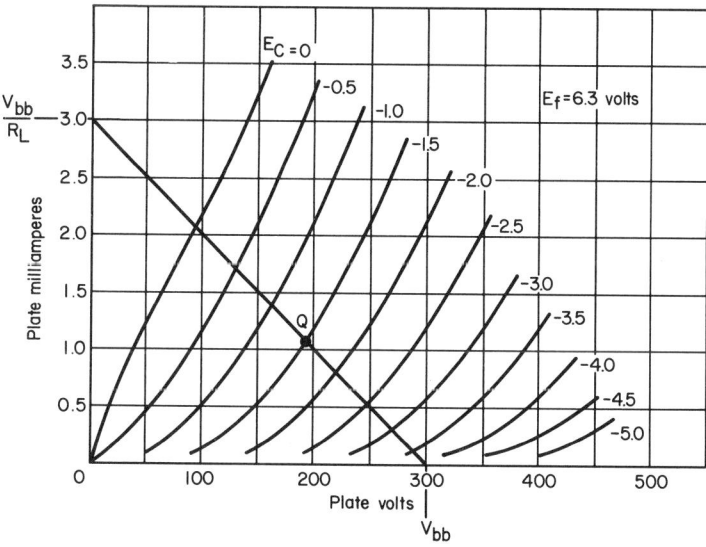

Fig. 3.9. Graphical vacuum-tube biasing.

58 Field-effect Transistors

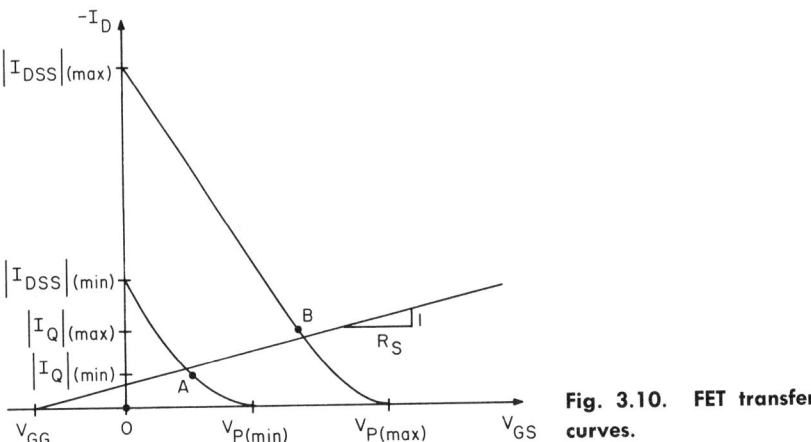

Fig. 3.10. FET transfer curves.

lege sophomore if need be; rather we will stick to graphic methods to illustrate the biasing of a single stage.

Most FET data sheets supply maximum and minimum values of I_{DSS} and V_P at room ambient, plus data to correct these quantities for temperature extremes. Where this information is not supplied, the designer is left to shift for himself. Let the graph in Fig. 3.10 represent temperature-corrected transfer curves for a hypothetical FET type. The curves do not show the temperature crossover point discussed in Chap. 2 because device selection variation is also included. Now suppose we want to design an amplifier whose quiescent drain current I_Q will not drift outside the limits $I_{Q(max)}$ and $I_{Q(min)}$ corresponding to the points A and B on the curves in Fig. 3.10. Changes of operating point on the transfer curve other than in a straight (constant voltage) vertical line imply that there is some external source resistance. When this resistance is plotted onto the transfer curves, its slope must keep the operating current greater than point A and less than point B. As shown in Fig. 3.10, the bias current limits determine the minimum value of self-biasing source resistance. They can also cause the load line to intersect the gate voltage axis at a point in the region of forward bias, as in Fig. 3.10, dictating the need for an external gate voltage that tries to forward-bias the FET.

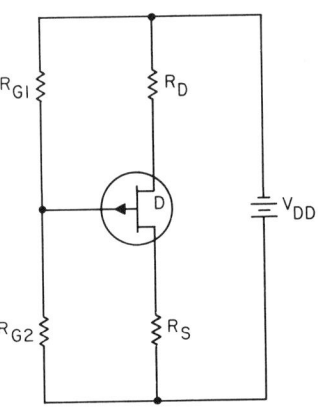

Fig. 3.11. FET amplifier with bias.

A circuit that will satisfy the conditions in the graph is given in Fig. 3.11. The FET amplifier with external gate bias voltage V_{GG} is obtained simply by using a resistive divider, $V_{GG} = aV_{DD}$, where a is the divider ratio. R_D is chosen to keep the drain-to-source voltage in the usable operating area of the output characteristics. This simple procedure is illustrated in Fig. 3.12. The usable operating area is bounded at low voltages by the voltage saturation characteristic, and at high voltages by the locus of points at which "apparent"

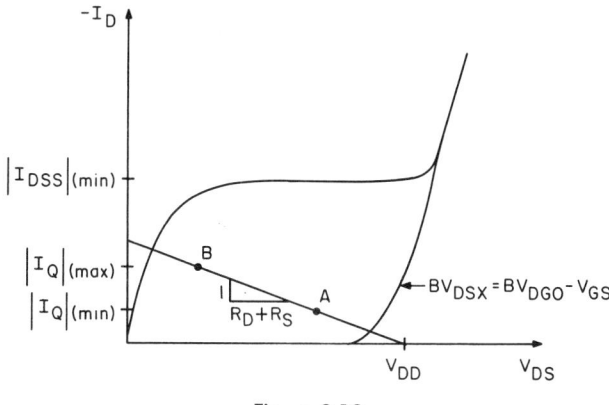

Figure 3.12

drain-to-source breakdown occurs. The example in Fig. 3.10 does not take into account the high-temperature variation of I_{GSX}, the gate leakage current. At moderate temperatures, the effect of gate leakage current in any gate bias resistor of practical size is negligible, but at elevated temperatures, the increase in I_{GSX} will produce a voltage drop across the equivalent gate resistance in the same sense as V_{GG}, i.e., toward forward bias. To compensate for this drop, both V_{GG} and R_S can be increased somewhat.

Generally the value of R_S necessary for stable biasing will be a significant part of the total drain-to-source loop resistance; this makes it necessary to use a source bypass capacitor to obtain high a-c gain.

3.3 DISTORTION IN FET AMPLIFIERS

The nature of the FET's square-law behavior permits both the harmonic and intermodulation distortion of a single-stage amplifier to be expressed in a rather simple closed form. By definition, the harmonic distortion produced by an amplifier whose input is a pure sinusoid is

$$D = \frac{\text{rms value of all harmonics in the output}}{\text{rms value of the fundamental of the output}}$$

Generally, input signals are not pure sinusoids but are complex music or speech waveforms which are composed of many frequencies and unique phase relationships. When two or more sine-wave signals are summed in a nonlinear network, beat frequencies are produced which are sums and differences of all integral multiples of the frequencies present in the input. The distortion produced by these beat frequencies is called intermodulation (I.M.) distortion and is defined as the rms value of the beat frequencies divided by the rms value of the complex input waveform. A standard method of measuring I.M. distortion is to apply two sine-wave signals, widely separated in frequency and of different amplitudes, to the amplifier input.

Harmonic Distortion. When a sinusoidal voltage is applied to the input of a common-source FET amplifier, the total instantaneous gate-to-source voltage can

Field-effect Transistors

be written as the sum of the quiescent bias voltage V_Q and the instantaneous input signal:

$$v_{GS} = V_Q + V_1 \sin \omega t \qquad (3.27)$$

From the square law, the total instantaneous drain current is

$$i_D = \frac{I_{DSS}}{V_P^2}(V_Q - V_P + V_1 \sin \omega t)^2 \qquad (3.28)$$

When Eq. (3.28) is expanded, it contains terms in $\sin \omega t$ and its second harmonic, as well as d-c terms:

$$i_D = \frac{I_{DSS}}{V_P^2}\left[(V_Q - V_P)^2 + \frac{V_1^2}{2} + 2(V_Q - V_P)V_1 \sin \omega t - \frac{V_1^2}{2}\cos 2\omega t\right] \qquad (3.29)$$

By definition, the harmonic distortion in i_D is

$$D = \frac{1}{4}\frac{V_1}{V_Q - V_P} \qquad (3.30)$$

Equation (3.30) can be written in more convenient form by solving I_{DP} for $(V_Q - V_P)$:

$$D = \frac{V_1}{4V_P}\sqrt{\frac{I_{DSS}}{I_Q}} \qquad (3.31)$$

If I_Q is allowed to go to zero, Eq. (3.31) blows up, because the square-law approximation of the transfer characteristic is double-valued over the i_D axis and mathematically acts as a full-wave rectifier (no fundamental) when $I_Q = 0$. For Eq. (3.31) to be of any use, it must be restricted to class A amplifiers.

Intermodulation Distortion. The standard method of measuring I.M. distortion of an amplifier consists of summing two sinusoids of known frequency and amplitude ratios at the input and measuring the resulting intermodulation components in the output signal.

$$i_D = \frac{I_{DSS}}{V_P^2}(V_Q - V_P + v_1 + v_2)^2 \qquad (3.32)$$

where
$$v_1 = V_A \sin \omega_1 t$$
$$v_2 = V_B \sin \omega_2 t$$

When Eq. (3.32) is expanded and fundamental identities are applied to v_1^2, v_2^2, and $v_1 v_2$, the resulting expansion contains both harmonic and intermodulation terms. The harmonic terms, of course, represent the harmonic distortion resulting from the nonlinear transfer characteristic operating on v_1 and v_2. Considering only the terms containing fundamental and intermodulation frequencies, we have

$$i_D = 2(V_Q - V_P)(V_A \sin \omega t + V_B \sin \omega t)$$
$$+ V_A V_B[\cos(\omega_1 - \omega_2)t - \cos(\omega_1 + \omega_2)t] \qquad (3.33)$$

To express the I.M. distortion as a ratio, the rms value of the intermodulation term in Eq. (3.33) is divided by the rms value of the fundamental:

$$\text{I.M.} = \frac{V_A V_B}{(V_Q - V_P)\sqrt{2(V_A^2 + V_B^2)}} \tag{3.34}$$

When expressed in terms of bias current I_Q, Eq. (3.34) becomes

$$\text{I.M.} = \frac{V_A V_B}{V_P \sqrt{2(V_A^2 + V_B^2)}} \sqrt{\frac{I_{DSS}}{I_Q}} \tag{3.35}$$

Effect of Unbypassed Source Resistance. Both Eqs. (3.31) and (3.35) can be rewritten in a form containing g_m in the numerator:

$$D = \frac{g_m V_1}{8 I_Q} \tag{3.36}$$

and

$$\text{I.M.} = \frac{g_m V_A V_B}{2 I_Q \sqrt{2(V_A^2 + V_B^2)}} \tag{3.37}$$

Now, in Eq. (3.8) we see that the effective transconductance of a common-source FET amplifier with an unbypassed source resistor is

$$g_{m'} = \frac{g_m}{1 + g_m R_S} \tag{3.38}$$

Thus the effect of unbypassed source resistance at a given bias point is to reduce distortion by $1/(1 + g_m R_S)$.

3.4 WIDEBAND AUDIO AND VIDEO AMPLIFIERS

RC-coupled Amplifiers. Although FET's are commonly found in amplifier circuits where they are used in conjunction with bipolar transistors, it is possible to use them exclusively in a cascaded, RC-coupled audio/video amplifier analogous to an all-vacuum-tube amplifier. Consider the RC-coupled amplifier in Fig. 3.13, which is an *n*-stage RC-coupled amplifier driven by a generator v_g.

At frequencies where the bulk resistance effects can be ignored, the behavior of the amplifier can be analyzed in much the same manner as a vacuum-tube

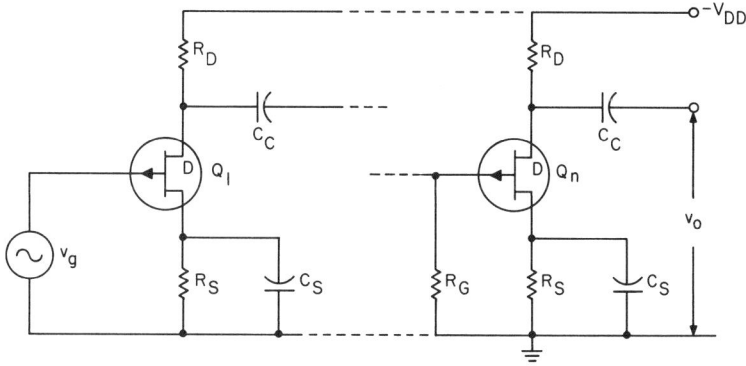

Figure 3.13

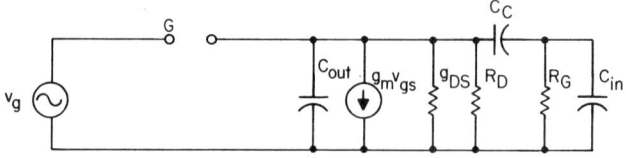

Figure 3.14

RC-coupled amplifier. Almost any text on electronic fundamentals will contain a detailed analysis of RC-coupled vacuum-tube amplifiers; so for our purpose it will be sufficient to present summarized results.

It is an almost impossible task to solve for the gain of a cascaded amplifier with a large number of stages by writing loop equations to describe the entire frequency range; the normal approach is to solve the gain of one stage at restricted frequency ranges. The equivalent circuit in Fig. 3.14 is terminated by the coupling capacitor C_C, the bias resistor R_G, and the input capacitance C_{in}, of the second stage. The frequency spectrum can be divided into three parts: the low frequencies, where the reactance of the coupling capacitor C_C is significant compared to R_G; a middle band of frequencies, where C_C can be considered as a short and the shunting capacitors as an open circuit; and the high-frequency range, where the loading of the shunt capacitances C_{out} and C_{in} becomes significant.

Implied in the above groupings is the assumption that the reactance of C_S is negligible to frequencies much lower than those where the reactance of C_C becomes significant. This may not always be the case, however. Actually, where it is desirable to connect a feedback loop around more than one stage, it may be advantageous to allow C_S to appear to "open up" at a much higher frequency than C_C. However, to keep this discussion simple, we will stick to our original assumption about C_S.

At the middle band of frequencies, the voltage gain of the first stage is simply

$$A_1 = g_m R_P \tag{3.39}$$

$$\text{where } R_P = \frac{R_D R_G}{R_D + R_G + g_{DS} R_D R_G} \tag{3.40}$$

The voltage gain of an *n*-stage amplifier is the product of the individual stages:

$$A_T = A_1 A_2 \ldots A_n \tag{3.41}$$

If the stages are identical, the gain is A^n.

At the low frequencies, the voltage gain is a function of the coupling capacitor, C_C, hence

$$A_{1Lo} = \frac{g_m R_P}{1 - j\frac{f_1}{f}} \tag{3.42}$$

Where f is the lower cutoff frequency or "break point" at which the reactance of C_C is equal to the loop resistance,

FET'S in Low-level Linear Circuits

$$R_G + \frac{R_D}{1 + R_D g_{DS}}$$

$$f_1 = \frac{1}{2\pi R_{loop} C_C} \tag{3.43}$$

Generally, R_G is very large compared to $R_D/(1 + R_D g_{DS})$.

At the high frequencies, the voltage gain can be approximated by

$$A_{1Hi} = \frac{g_m R_P}{1 + j\dfrac{f}{f_2}} \tag{3.44}$$

where the upper cutoff frequency f_2 is given by

$$f_2 = \frac{1}{2\pi R_P (C_{out} + C_{in})} \tag{3.45}$$

A graph of the voltage gain vs. frequency is shown in Fig. 3.15. At the break points f_1 and f_2, the gain is down 3 db from the maximum or middle frequency gain, which is normalized to 0 db in the graph. There are no second-order terms in the above expressions for voltage gain; therefore, the "rate of fall" of the curve in Fig. 3.15 is asymptotic to 6 db/octave at both ends of the spectrum.

When f_1 is very small compared to f_2, the overall bandwidth of the stage can be approximated by f_2. If all n stages of the amplifier in Fig. 3.13 are identical, the cutoff frequency is

$$f_{2n} = f_2 \sqrt{2^{1/n} - 1} \tag{3.46}$$

The gain-bandwidth figure of merit for an FET is obtained by multiplying Eq. (3.39) by Eq. (3.45), which results in

$$A_1 \Delta f = \frac{g_m}{2\pi (C_{out} + C_{in})} \tag{3.47}$$

From Eq. (3.11), the lowest possible value of the input capacitance occurs when

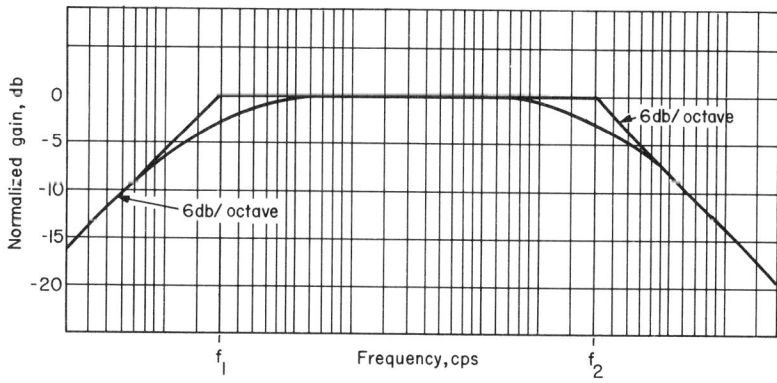

Fig. 3.15. Frequency response of first stage.

64   Field-effect Transistors

$A_2 = +1$; also, C_{DG} is a good approximation to the output capacitance of the first stage, then

$$A_1 \Delta f \approx \frac{g_m}{2\pi(C_{DG} + C_{SG})} = \frac{g_m}{2\pi C_{is}} \tag{3.48}$$

The gain-bandwidth product is a kind of cutoff frequency. It can be thought of as the cutoff frequency of an amplifier that uses some feedback technique to give it unity gain at mid-band. An FET amplifier with a voltage gain of one is still quite useful as an impedance transformer. A 2N2497, as an example, with a typical g_m of 1,500 μohms and a typical C_{is} of 15 pf, will be useful to about 15 mc. Up to this frequency one can trade gain for bandwidth, if one is aware of the danger that lurk in allowing the break frequencies of several stages to occur at too close frequency intervals.

An example of an RC-coupled FET amplifier is given in Fig. 3.16. This amplifier was used to drive high-impedance headphones in an optical communication system; it served the dual purpose of amplification and frequency compensation. The series peaking capacitors C_A and C_B compensate for high-frequency inadequacies in the rest of the system. One advantage of this amplifier is that it has no large electrolytic coupling and bypass capacitors. Without the peaking capacitors, the amplifier voltage gain is about 400 (depending somewhat on device selection) and the upper and lower break frequencies are 17 cps and 35 kc respectively. No attempt was made to stagger the break frequencies of the individual stages; so it would be unwise to try to connect a feedback loop around this amplifier.

Wideband Amplifiers Using FET-Bipolar Transistor Combinations. Since the outstanding low-level characteristic of FET's are high input impedance and low noise, FET's are naturally most useful at very-low-level, high-impedance points in elec-

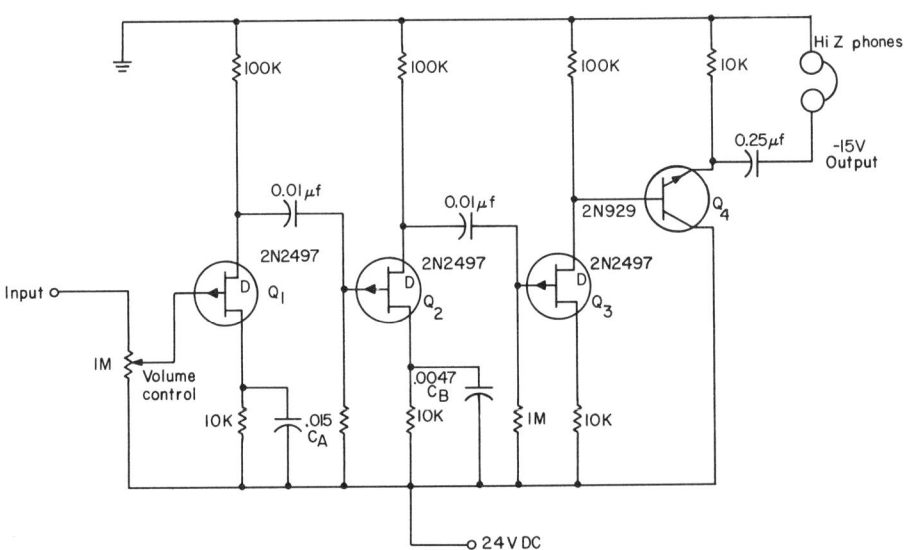

Fig. 3.16. An RC-coupled amplifier.

FET's in Low-level Linear Circuits

tronic circuits, e.g., as the load on a high-internal-impedance transducer. Once the impedance level is reduced to where conventional or bipolar transistors can be driven, it may not be economical to continue using FET's for further amplification. Using FET's and bipolar transistors in combination offers other advantages that are not readily apparent, as we shall soon discover.

The low breakdown voltages found in FET's make it difficult to realize near unity gain in a source-follower. For example, the g_m of an FET will be 1,000 μmhos at about 1 ma of drain current; to obtain a voltage gain of 0.98, R_S must be greater than 50 K, and the supply voltage must exceed 50 volts by an amount somewhat larger than the pinch-off voltage. To avoid having to use such a high supply voltage (in most applications it will simply not be available) we might consider replacing the source bias resistor R_S with a dynamic constant-current supply.

Two possibilities are shown in Figs. 3.17a and b. In Fig. 3.17a, a PNP transistor serves as the constant-current supply. Its base is connected to a fairly "stiff" voltage divider R_1 and R_2, and the bias current is controlled by the value of R_E. The transistor Q_2 has a very high dynamic output resistance in common base, perhaps several hundred megohms. If the external load is light, the voltage gain approaches the $\mu/(\mu + 1)$ of the FET, where $\mu = g_m/g_{DS}$. In Fig. 3.17b an FET is used for the constant-current supply thus eliminating two resistors; the bias current is controlled by varying R_S.

A variation of this constant-current biasing technique which can give voltage gains greater than one is the basic compound connection shown in Fig. 3.18. The circuit has been called a "bootstrapped source-follower." The drain of the FET drives the base of the NPN transistor, whose collector drives the source of the FET in phase with the input signal. With the equivalent circuit in Fig. 3.19 we can quickly analyze the circuit for its low-frequency performance. The input resistance

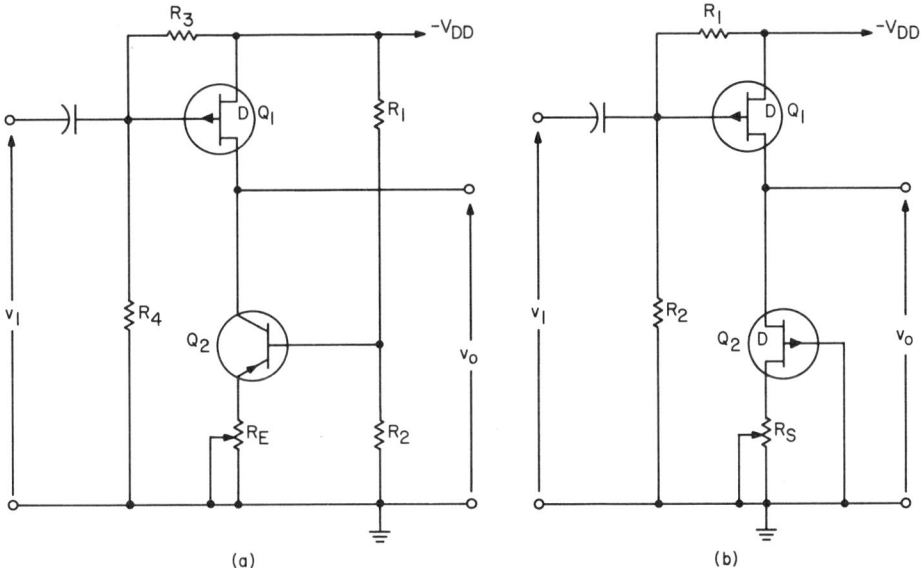

Fig. 3.17. Source-follower with constant-current supply.

66 Field-effect Transistors

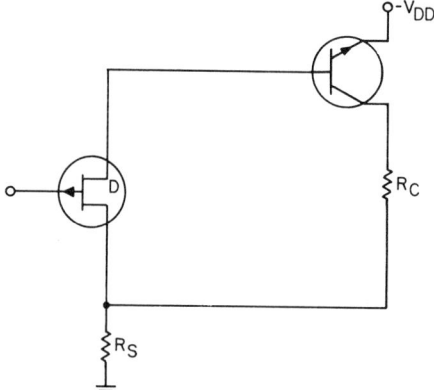

Fig. 3.18. The bootstrapped source-follower.

of the transistor is assumed to be negligible; therefore, the performance equations of the circuit simplify to

$$g_m v_{gs} + h_{fe} i_b = v_s \left(g_{DS} + \frac{1}{R_S} \right) \quad (3.49)$$

$$v_{gs} = v_1 - v_s \quad (3.50)$$

$$i_b = g_m v_{gs} - v_s g_{DS} \quad (3.51)$$

and
$$v_o = v_s + h_{fe} i_b R_C \quad (3.52)$$

A few algebraic manipulations and we find that

$$\frac{v_s}{v_1} = \frac{g_m(1 + h_{fe})R_S}{1 + (g_{DS} + g_m)(1 + h_{fe})R_S} \quad (3.53)$$

$$\frac{v_o}{v_1} = \frac{g_m[(1 + h_{fe})R_S + h_{fe}R_C]}{1 + (g_{DS} + g_m)(1 + h_{fe})R_S} \quad (3.54)$$

Referring back to Eq. (3.20) we see in Eq. (3.53) that the source resistor appears to be multiplied by $(1 + h_{fe})$ and the gain approaches $g_m/(g_{DS} + g_m)$. This is good also for distortion reduction. Equation (3.54) reduces to approximately $[g_m/(g_m + g_{DS})](R_S + R_C/R_S)$ for large values of h_{fe}, or roughly the ratio $R_S + R_C/R_S$.

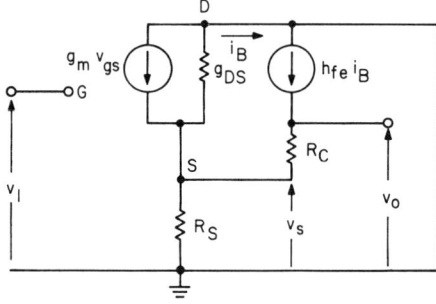

Figure 3.19

FET'S in Low-level Linear Circuits   67

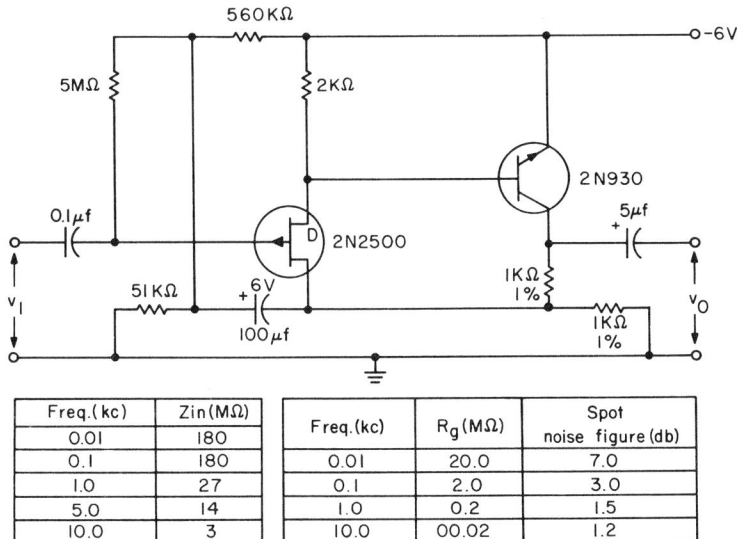

Freq.(kc)	Zin(MΩ)
0.01	180
0.1	180
1.0	27
5.0	14
10.0	3

Freq.(kc)	R_g (MΩ)	Spot noise figure (db)
0.01	20.0	7.0
0.1	2.0	3.0
1.0	0.2	1.5
10.0	00.02	1.2

Fig. 3.20. 6-db transducer amplifier with performance data. Noise-figure measurements were made with optimum generator resistance at each frequency.

Figure 3.20 shows a bootstrapped source-follower design with a 6-db gain intended for high-impedance transducer outputs. In this circuit, the 5-megohm gate bias resistor is bootstrapped to the source through the large electrolytic capacitor (100 mfd). The bias resistor is effectively multiplied by $1/(1 - v_s/v_1)$; hence the very large input resistances at low frequencies in the accompanying table.

We have seen circuit techniques for approaching the theoretical maximum source-follower gain at low frequencies, thus according to Eq. (3.23) virtually eliminating C_{SG}, but the drain-to-gate capacitance C_{DG} has been unaffected. This capacitance can also be bootstrapped by a connection like that in Fig. 3.21. A resistor is connected between the drain of Q and the supply; the drain is driven in phase with the input signal, effectively reducing C_{DG} by $(1 - A_1 A_3)$, where A_1 is the source-follower gain of Q_1, and A_3 is the emitter-follower gain of Q_3. The source is supplied with a constant current from Q_2, ensuring a high source-follower gain.

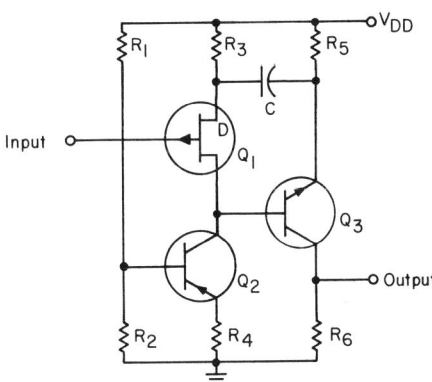

Fig. 3.21. Bootstrapping the drain.

68   Field-effect Transistors

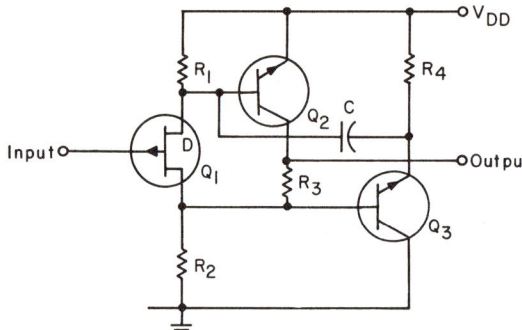

Fig. 3.22. Both terminals bootstrapped.

A variation of the circuit in Fig. 3.21 is shown in Fig. 3.22. Here, both the drain and the source are bootstrapped! The techniques in Figs. 3.21 and 3.22 can reduce the effect of the FET capacitances to the point where they are insignificant compared to the stray circuit capacitances at the input terminal. The stray capacitances can be all but eliminated by enclosing the input stage of the amplifier in a conducting shield and connecting the shield to a point of near-unity voltage gain, e.g., through a large capacitor to the source of Q_1 in both Figs. 3.21 and 3.22.

A design example utilizing the technique of Fig. 3.21 is shown in Fig. 3.23. An extra stage, Q_4, is added to give some gain to compensate for the fact that the source-follower and emitter-follower gains are not quite unity. The 10-K pot provides the compensation adjustment. One must be careful with this adjustment as it can cause instability (i.e., oscillation). The surest way to obtain optimum adjustment is to apply a 10-kc square wave to the input and adjust the 10-K pot for minimum overshoot on the output waveform when displayed on a properly compensated wideband oscilloscope. The frequency response curves in Fig. 3.24 were plotted with the 10-K pot peaking adjustment set to zero. The dashed curve for $R_g = 10$ megohms shows how the response was improved by adjusting the overshoot on the square-wave response. The voltage gain of this amplifier is about 6 db at low frequencies.

Another interesting compound connection of an FET and a bipolar transistor is the direct-coupled cascode circuit in Fig. 3.25. A common-source FET, Q_1 drives a common-base transistor, and the base supply voltage V_{BB} establishes a drain-bias voltage for Q_1. This connection features a large amount of isolation between the output and the input; i.e., the reverse transfer admittance is small, suiting it for high-frequency tuned amplifiers as well as video amplifiers. Another advantage is that the transistor stands the drain supply voltage V_{DD} while the FET sees only V_{BB}; transistors can be found with common-base breakdown voltages exceeding 100 volts. This would allow the FET to be biased at a fairly high drain current for high g_m; a large collector load resistor R_2 can then be connected to Q_2 for high voltage gain if a supply voltage is available that is in the order of the collector-base breakdown voltage of the transistor. In some cases it may be possible to use this circuit as a direct vacuum-tube replacement, particularly if an N-channel FET and an NPN transistor are used. In such a case, the polarities of the supply voltages of Fig. 3.25 would be reversed.

FET's in Low-level Linear Circuits

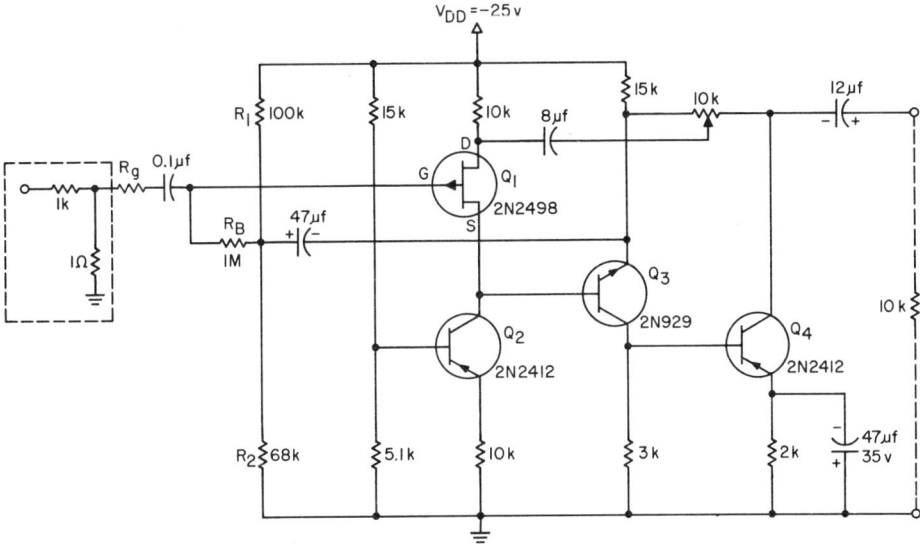

Fig. 3.23. 6-db high-input z FET amplifier.

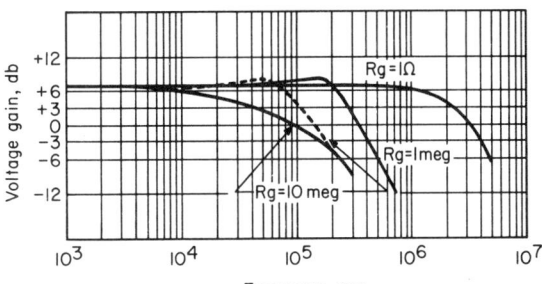

Fig. 3.24. FET amplifier response curves.

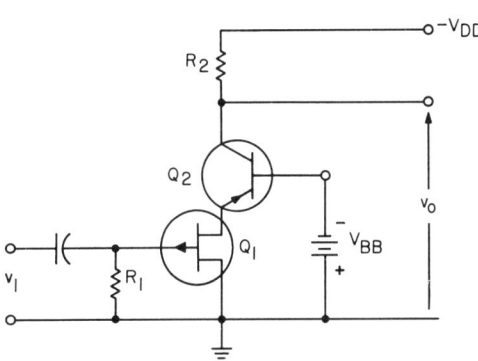

Fig. 3.25. Direct-coupled cascode circuit.

3.5 DIRECT-COUPLED AMPLIFIERS

The simplest form of an FET d-c amplifier is the impedance-transforming source-follower in Fig. 3.26. This circuit can be made quite insensitive to changes in ambient temperature by selecting the proper bias point according to the method outlined in Chap. 2; that is, R_S must be selected so that the drain current is at the temperature-stable point for a given FET. When the FET is properly biased, the load line will intersect at the stable point as shown in Fig. 3.27. Of course it is obvious that R_S will have to be readjusted every time the device is changed, but this type of circuit can give extremely stable operation over a wide temperature range provided that d-c signals are kept small and the generator resistance is reasonably low. High values of gate resistances lead to instability at high temperatures because of the voltage drop caused by the gate leakage current I_{GSX}, which shifts the load line in Fig. 3.27 to the left.

The best way to compensate for the drift caused by gate leakage current is to use a matched pair of FET's in a direct-coupled difference amplifier. The analysis of an FET difference amplifier in terms of thermal drift follows from the simplified equivalent circuit in Fig. 3.28. Each FET can be represented by an ideal "driftless" amplifier with equivalent input drift current and voltage generators, the same as the noise representation (drift is a kind of noise). It is common practice to express the performance of a d-c amplifier in terms of its total equivalent input drift voltage; this is the input voltage change needed to maintain a constant output voltage under varying ambient temperatures. In Fig. 3.28, the equivalent input drift is $\Delta v_{in}/\Delta T$, $\Delta v_o/\Delta T = 0$ by definition.

Writing Kirchhoff's voltage around the input loop,

$$\frac{\Delta v_{in}}{\Delta T} = \frac{\Delta(V_{GS1} - V_{GS2})}{\Delta T} + \frac{\Delta(I_{G1} - I_{G2})}{\Delta T} R_G \qquad (3.55)$$

Thus we see the obvious, that both $\Delta V_{GS}/\Delta T$ and $\Delta I_G/\Delta T$ should be matched between the two FET's for minimum drift. If the two FET's were individually adjusted for the zero drift point, $\Delta(V_{GS1} - V_{GS2})/\Delta T$ would be zero. However, we saw in Fig. 2.6 that $\Delta V_{GS}/\Delta T$ for a given FET is completely specified by I_{DSS} and V_P. Matching these two quantities will ensure minimum drift at any bias point. The two FET's do not have to be individually adjusted, nor do any adjustments have to be made when the devices are changed. The only requirement is a matched pair.

The leakage current is so small that it is negligible at moderate ambient temperatures; it becomes significant only at high temperatures. It seems likely, then, that matching at a single high worst-case temperature will yield maximum information for a minimum of effort. Consider Fig. 3.29, a plot of gate leakage vs. temperature for devices. The two curves

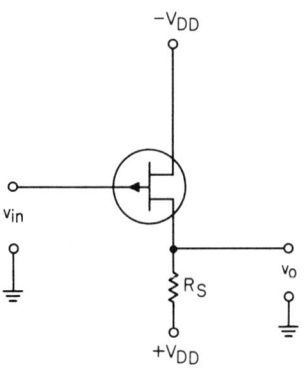

Fig. 3.26. Simple d-c impedance transformer.

FET'S in Low-level Linear Circuits   71

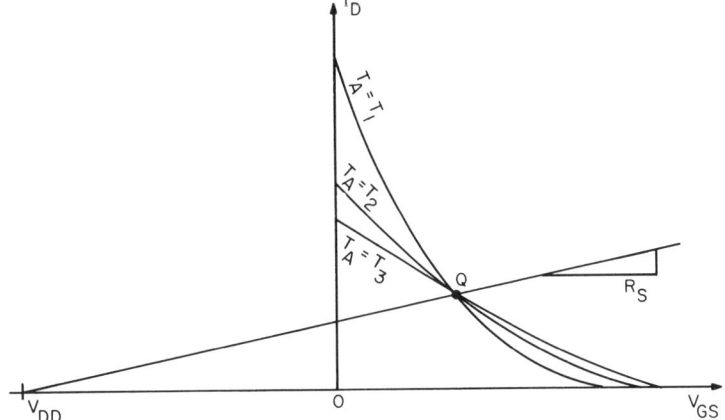

Fig. 3.27. The bias load line intersects the stable point.

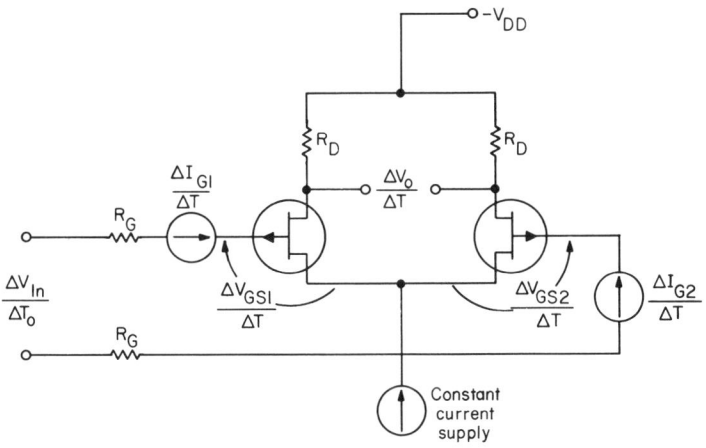

Fig. 3.28. D-c difference-amplifier drift equivalent.

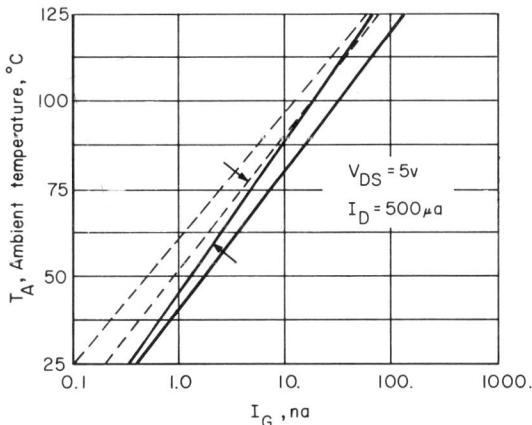

Fig. 3.29. Matching leakage currents.

72   Field-effect Transistors

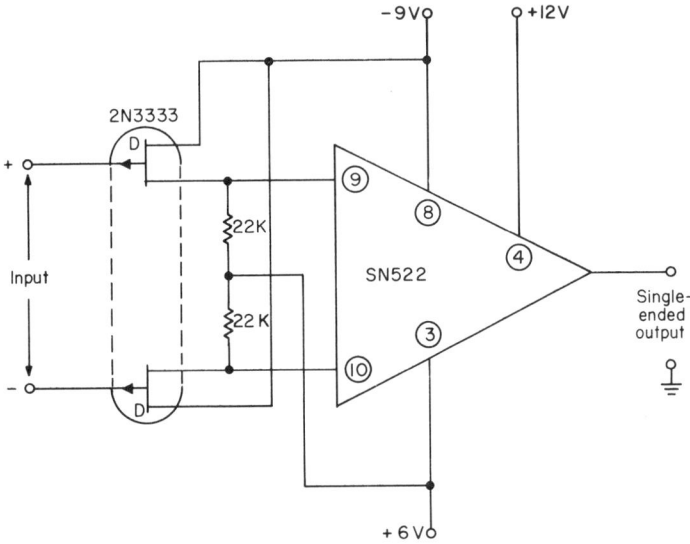

Fig. 3.30. A high-impedance integrated-circuit operational amplifier.

indicated by the arrows make the best match even though they do not track very well at lower temperatures.

Conventional bipolar transistors perform much better as d-c difference amplifiers than do FET's when the generator impedances are below about 500 kilohms. They generally can be matched to a much tighter equivalent voltage drift and their equivalent input noise voltage does not exhibit the large $1/f$ component as do the FET's. Normal base currents, however, are several orders of magnitude higher than FET leakage currents, even when the transistor is operated with its emitter current in the low microampere range. Therefore, FET's are much superior to bipolar transistors in d-c difference amplifiers when driven with generator resistances in the megohm and tens of megohms range.

The 2N3333-36 series of FET's are matched pairs in a single integrated-circuit-type package. This matched pair is very useful for converting integrated-circuit-type operational amplifiers to high-impedance inputs as in Fig. 3.30. The SN522 is a differential input operational amplifier which normally has an input impedance of about 15 K.

3.6 FET'S AND HIGH FIDELITY

In the past, the possibility of using FET's in high-fidelity circuits excited the curiosity of many electronic engineers who were audio hobbyists, but they were disappointed when they saw then-current FET price tags. This state of affairs will not always be; at the rate FET prices are tumbling, they are fast becoming quite competitive for a good many consumer applications. Of particular interest is the possibility of encapsulating a dual FET like the 2N3336 in an inexpensive epoxy package, which would make its price about equal to the price of a 12AX7. This dual device would not require any of the matching specifications of the

2N3336; it would require just the limits on I_{DSS} and V_P, plus good low-leakage and low-noise specifications.

High-fidelity preamplifiers employing FET's are considerably less complex than transistor preamplifiers, require fewer devices (because of the high power gain of FET's), and less complicated biasing. In fact, some vacuum-tube preamplifiers have been known to work satisfactorily when FET's were substituted directly for the vacuum tubes by simply lowering the supply voltage, and when P-channel FET's were used, reversing the polarity of the supply. The devices needed to perform this trick are not yet commercially available but are in the developmental stage. What is required is a device with the general specifications of the 2N3336 and a breakdown voltage about 50 per cent higher.

The author's vacuum-tube preamplifier was converted to FET's with excellent results. In order to use commercial devices, it was necessary to extend the breakdown voltage by using the cascode connection of Fig. 3.25. The PNP transistor used is a 2N398B, germanium alloy and relatively inexpensive. Eight of these circuits were used, two each in a small module with a nine-pin tube socket adaptor, to replace four 12AX7's. The preamplifier was made by Dyna and had given about five years of faithful service during its vacuum-tube period, and it was with much hesitation that the existing power supply was torn out and replaced by a negative supply of -150 volts. The d-c supply for the transistor bases is obtained with a 10-volt reference diode and a large dropping resistor (150 K) off the -150-volt supply. Figure 3.31 is the schematic of one channel of the preamplifier when FETs have been substituted for vacuum tubes.

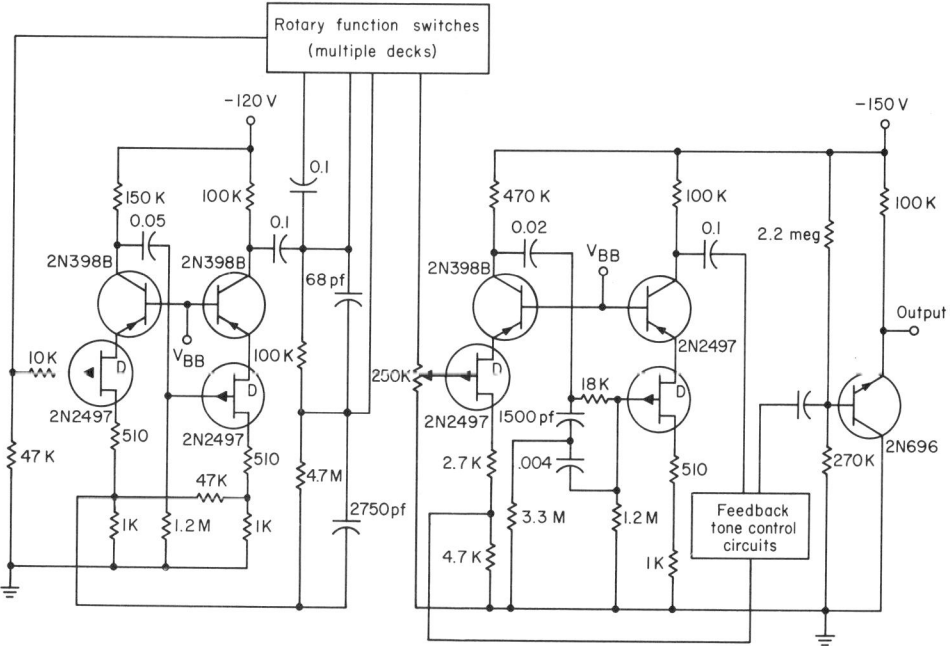

Fig. 3.31. Modified vacuum-tube preamplifier.

74 Field-effect Transistors

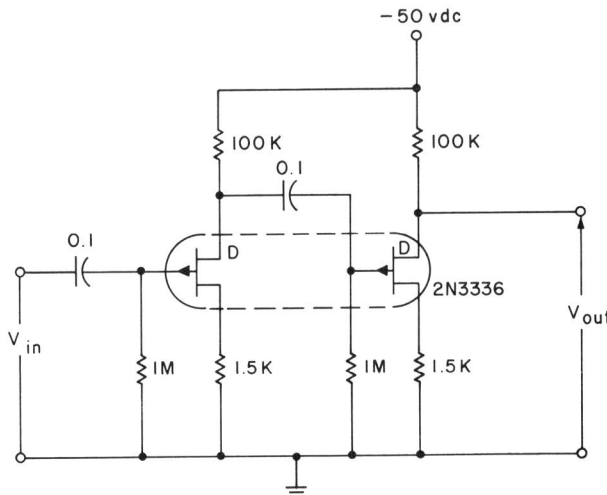

Fig. 3.32. Building-block amplifier.

The extra bias resistors in the FET sources were built into the plug-in modules. These resistors compensate for differences between the bias characteristics of the 2N2497 and the 12AX7. The emitter-followers match the preamplifier output to the relatively low input impedance of a transistor power amplifier. One problem cropped up with the germanium alloy transistors; since some of them exhibit excessive $1/f$ current noise, they had to be specially selected. The modified preamplifier meets Dyna's original specifications regarding noise, distortion, and frequency response; the added feature is that the filament power has been considerably reduced. Because of the lowered supply voltage, the output voltage at 20 cps before clipping is only 2.5 volts rms compared to about 10 volts for the vacuum-tube preamplifier, but since the power amplifier being driven will put out its full 70 watts with 1.2 volts rms input, there is no problem.

Admittedly the circuit is complicated by the necessity for the PNP transistors, but if devices like the 2N3336 were available at a commercially competitive price, a much simpler circuit could be designed. The preamplifier shown in Fig. 3.31 is composed of two "building-block" amplifiers of two stages each. One of the blocks performs the equalization function and the other contains the tone controls; the preamplifier is designed around the building blocks. A building-block amplifier using the 2N3336 is shown in Fig. 3.32. Equalization and tone-control networks can be designed around two circuits like this, requiring just two "packages" for the whole preamplifier. This circuit has a voltage gain of 1,000 and a gain-bandwidth product of 15 mc.

Another audio application for which FET's are ideally suited is preamplifiers for condenser microphones. The two-terminal equivalent circuits of a condenser microphone consist of a voltage generator and a small series capacitor, usually 50 to 200 pf. To get good low-frequency response out of this, an amplifier with a high input impedance is required—the requirement cries out for FET's. Because of the small size and low power requirements of an FET, the preamplifier can

FET'S in Low-level Linear Circuits 75

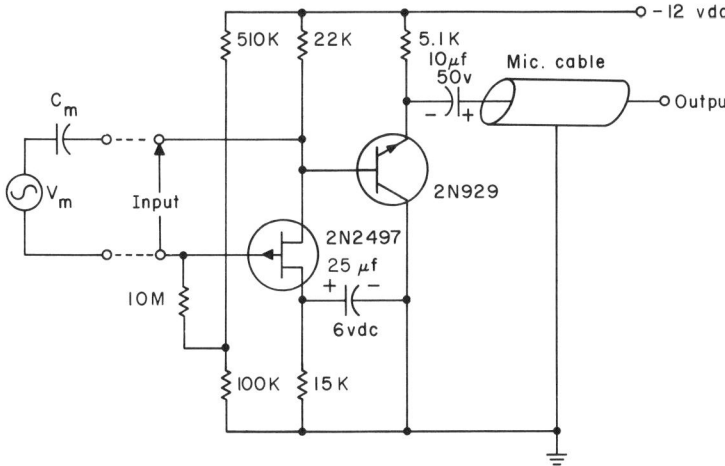

Fig. 3.33. Condenser microphone preamplifier.

easily be mounted in the microphone itself. A condenser microphone preamplifier that works very well with the small capacitance input is shown in Fig. 3.33.

The microphone input is applied between the drain and the gate of the input FET; this puts the coupling capacitor in a shunt feedback path and effectively multiplies it by one plus the voltage gain of the stage, taking ground potential as a reference. The capacitance of the microphone cable is driven by an emitter-follower whose output impedance is about 100 ohms; this will drive 500 feet of cable without appreciably affecting the frequency response. Frequency response and harmonic distortion curves for the preamplifiers are given in Fig. 3.34. The microphone was simulated by the secondary of an interstage transformer with a 100-pf coupling capacitor in series. The amplifier output level was 100 mv rms at 1 kc.

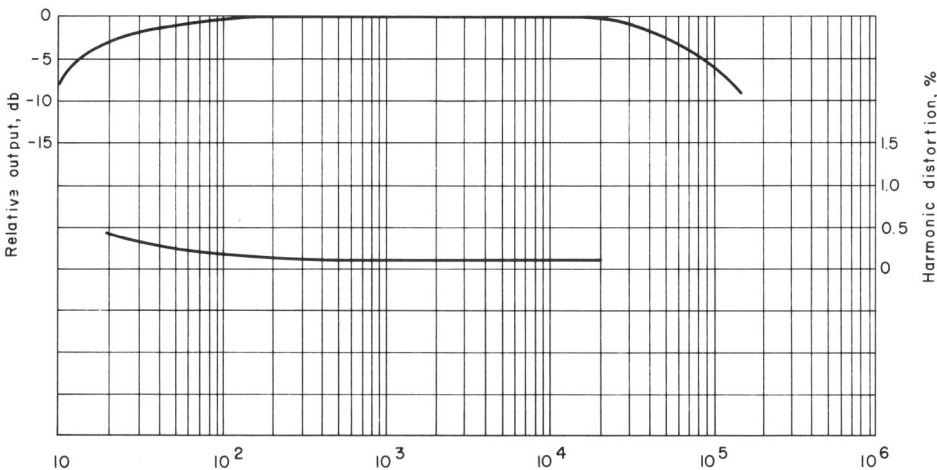

Fig. 3.34. Condenser-microphone preamplifier performance.

76 Field-effect Transistors

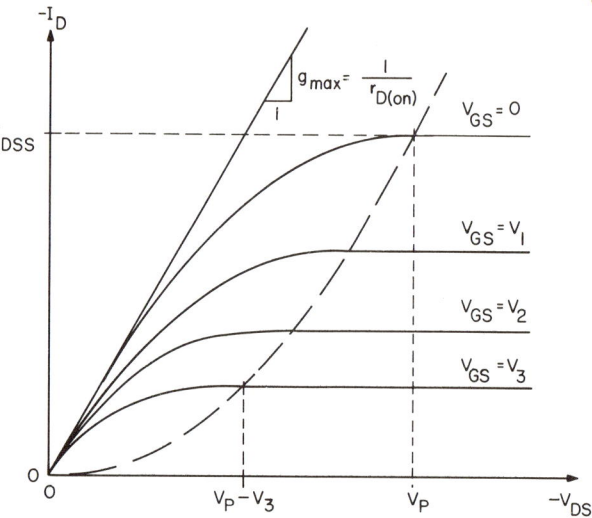

Fig. 3.35. Output characteristics in triode region.

3.7 AUTOMATIC GAIN CONTROL

A set of FET output curves is shown in Fig. 3.35 with the drain-to-source voltage axis expanded to show the triode regions in detail. In Chap. 1 we noted the interesting fact that the zero-current channel resistance at any gate bias voltage is equal to the reciprocal of the pinched-off transconductance at the same bias. Thus in Fig. 3.35 the slope of the output curve at $V_{GS} = 0$ is equal to $1/r_{D(on)}$ and it is also equal to g_{max}. If the FET described by these curves is a square-law device, the slope intersects the dotted constant-current line drawn from I_{DSS} at $V_P/2$ (remember that $g_{max} = 2I_{DSS}/V_P$). The simplest form an automatic gain control (AGC) circuit would take is the variable-voltage divider shown in Fig. 3.36. When the FET is pinched off, $V_{AGC} > V_P$, and the transmission of the network depends only on the series resistor R_S and the loading at the output. If the load can be neglected, $v_o/v_{in} = 1$. When V_{AGC} is less than V_P the attenuation of the network is given by

$$\frac{v_o}{v_{in}} = \frac{r_{DS}}{R_S + r_{DS}} = \frac{1}{1 + g_m R_S} \qquad (3.56)$$

Fig. 3.36. Simple AGC circuit.

Assuming square-law behavior for the FET, we have

$$\frac{v_o}{v_{in}} = \frac{1}{1 + \frac{2I_{DSS}}{V_P}\left(\frac{V_{AGC}}{V_P} - 1\right)R_S} \quad (3.57)$$

The minimum possible value of Eq. (3.57) occurs when $V_{GS} = -\phi$, thus

$$\left(\frac{v_o}{v_{in}}\right)_{min} = \frac{1}{1 - \frac{2I_{DSS}}{V_P}\left(\frac{V_P + \phi}{V_P}\right)R_S} \quad (3.58)$$

It is possible to obtain a large AGC range this way; for example, if $R_S = 500$ K and the FET has the following characteristics:

$$I_{DSS} = 2 \text{ ma}$$
$$V_P = 2 \text{ volts}$$
$$\phi = 0.5 \text{ volt}$$

the maximum AGC range is the reciprocal of Eq. (3.58) or about 1250, and this is greater than 60 db. It is evident from Fig. 3.35 that the signals should be kept fairly small because the nonlinearity of the drain-to-source resistance can introduce large amounts of distortion. The situation is helped somewhat by the fact that the FET is most nonlinear for least attenuation (smallest input signals). This scheme has the advantage of simplicity, but it also has disadvantages; namely, the circuit has nc gain, it only attenuates, and the signal-to-noise ratio is worsened by a resistive divider. The signal-to-noise ratio degrades by 3 db every time the attenuation is doubled.

Figure 3.37 shows a circuit that uses an FET as a variable-emitter resistor in a common-emitter transistor amplifier. The variable resistor is the 2N2498. It is capacitance-coupled to the emitter, another FET. A low-current 2N3328 is used to supply a constant emitter-bias current, and to provide a very light dynamic loading on the emitter for maximum AGC range. The approximate expression for the gain of this circuit is

$$A = -\frac{R_C}{r_{DS}} = -g_m R_C \quad (3.59)$$

assuming that the 2N2498 is a square-law device,

$$A = -\frac{2I_{DSS}}{V_P}\left(\frac{V_{AGC}}{V_P} - 1\right)R_C \quad (3.60)$$

The voltage gain is a linear function of the AGC voltage, at least so long as the FET transfer curve is a parabola. The voltage gain of the circuit in Fig. 3.37 is shown plotted against the AGC voltage in Fig. 3.38. The curve is linear over part of the range but tends to "remote" at high AGC voltages. This is due to the FET's departure from square-law behavior at low drain currents. Another FET, a 2N3336 type device, was substituted in the circuit to see if its "remoting" char-

78 Field-effect Transistors

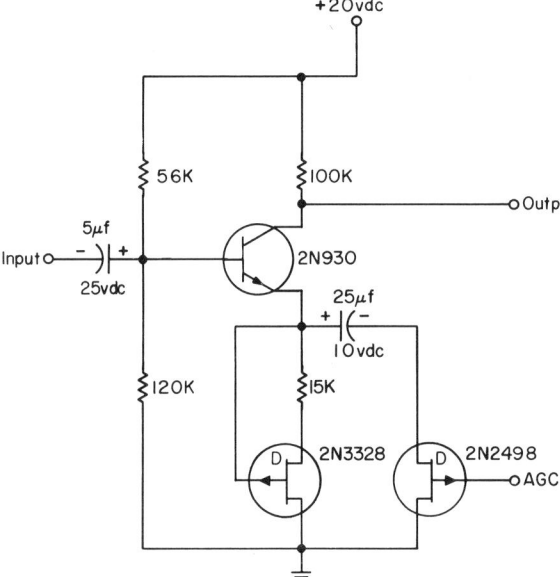

Fig. 3.37. Amplifier stage with provision for AGC.

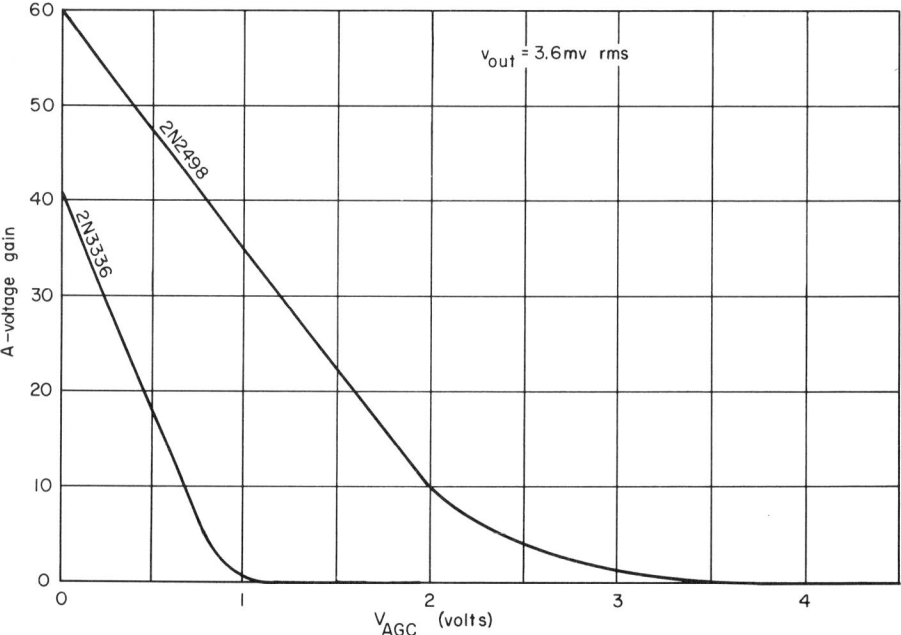

Fig. 3.38. Voltage gain vs. control voltage.

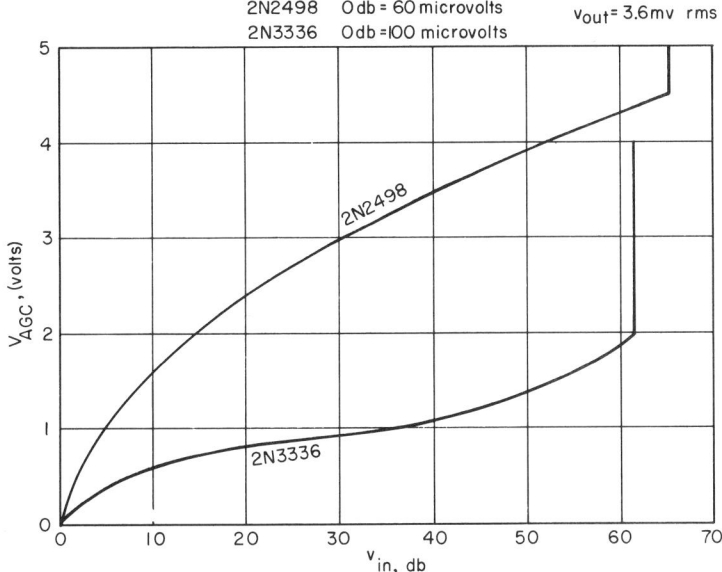

Fig. 3.39. AGC range.

acteristic was different, but the same tendency is shown. If both devices cut off sharply, that is, if they followed square-law behavior right down to zero drain current, the curves would intersect the $A = 0$ axis at $V_P = 2I_{DSS}/g_{max}$. However, the true pinch-off voltages of these devices are quite a bit higher than those predicted by the square law. Unfortunately, since most of the gain control of the amplifier lies in the "remoting" region, the plots in Fig. 3.38 do not contain much useful information. A better look at the gain-control range can be taken by plotting the AGC voltage vs. the input voltage, for a constant output; this is done in Fig. 3.39. With either device as the variable resistor, the circuit has an AGC range greater than 60 db. The maximum input voltage in both cases is about 100 mv.

Since the variable resistor is capacitor-coupled to the emitter of the transistor, there is no change in bias current when a strong AGC voltage is suddenly applied. Thus, no transient "thump" intrudes in the amplifier output voltage. This fact makes the circuit desirable for broadcast speech compressors.

4

FET's in Nonlinear Circuits

The largest area of application for nonlinear circuits is in digital systems. In competing with bipolar transistors in this area, the FET has some inherent limitations, such as high ON or *saturation* resistance and relatively slow speed. The fastest of currently available FET's can only operate at one-twentieth the speed of some transistors. For these reasons FET's rarely displace bipolar transistors from switching circuits, and there is little hope that they will in the future.

Any advantages FET's may have derive from their high input impedance at low frequencies and the fact that the FET is a majority-carrier device. The high-impedance level allows large fan-outs in logic circuits. We assume that the mechanism of channel conduction (majority carriers only) should make the FET considerably less sensitive to radiation damage than bipolar transistors, but this has not yet been proved conclusively.

An interesting, if rather limited, application area to which FET's seem ideally suited is the area of voltage-squared circuits. By taking advantage of the square-law behavior of some FET's it is possible to design circuits that square voltages to a high degree of accuracy; this will be discussed in some detail.

4.1 FET SQUARING CIRCUITS

Accurate square-law devices or squaring circuits are used mainly in the fields of noise investigation, analog computation, and measurement of power in complex waveforms. In these applications, accuracy and bandwidth are important factors. Noise investigations require a squarer of high accuracy, and often, wide bandwidth. In general, computing work requires a high order of accuracy and, if possible, a wide bandwidth; however, bandwidth must often be sacrificed to achieve the desired accuracy since these quantities are to some extent inversely related.

The square-law approximation to the pinch-off drain current,

$$i_{DP} = I_{DSS}\left(\frac{v_{GS}}{V_P} - 1\right)^2 \tag{4.1}$$

contains zero-, first-, and second-order terms in v_{GS}. If two FET's were to be connected as in Fig. 4.1 with the drains tied together and their inputs fed out of phase,

FET's in Nonlinear Circuits

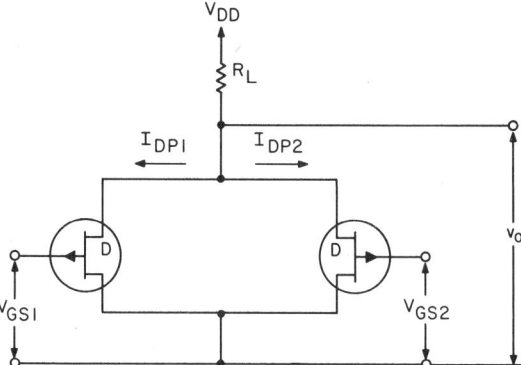

Fig. 4.1. Basic FET squaring circuit.

the first-order terms would tend to cancel in the load resistor R_L. The total gate-to-source voltage of each device is the sum of the quiescent voltage V_Q and a varying signal v_{in}. Thus, if the signals to the two gates in Fig. 4.1 are fed out of phase, we have

$$v_{GS1} = V_{Q1} + v_{in} \tag{4.2}$$

$$v_{GS2} = V_{Q2} - v_{in} \tag{4.3}$$

The output voltage is produced by the current that flows in R_L, which is the sum of the two drain currents. Substituting Eqs. (4.2) and (4.3) into Eq. (4.1) for each drain current and summing the two, we obtain the current in R_L. Assuming the two FET's are matched for I_{DSS} and V_P,

$$\begin{aligned}
i_L &= i_{DP1} + i_{DP2} \\
&= I_{DSS}\left[\left(\frac{V_Q + v_{in}}{V_P} - 1\right)^2 + \left(\frac{V_Q - v_{in}}{V_P} - 1\right)^2\right] \\
&= \frac{2I_{DSS}}{V_P^2}(V_Q^2 - 2V_QV_P + V_P^2 + v_{in}^2) \\
&= \frac{2I_{DSS}}{V_P^2}(V_Q - V_P)^2 + \frac{2I_{DSS}}{V_P^2}v_{in}^2 \tag{4.4}
\end{aligned}$$

Thus if the two FET's are identically matched, the first-order terms cancel in the load resistor and the zero- and second-order terms add. In general, for a transfer curve expressible by an n-order power series, the odd-order terms cancel and the even-order terms add. If only a-c operation is desired, the output may be coupled through a capacitor, eliminating the d-c term, and the a-c output voltage will be

$$v_o = \frac{2I_{DSS}R_L}{V_P^2}v_{in}^2 \tag{4.5}$$

If d-c operation is desired, it would be to our great advantage to try to bias the two FET's at their temperature-stable point. The operation of the circuit is illustrated graphically in Fig. 4.2. The two transfer-characteristic curves of the matched transistors cross at some common quiescent point. The sum of the two forms a

82   **Field-effect Transistors**

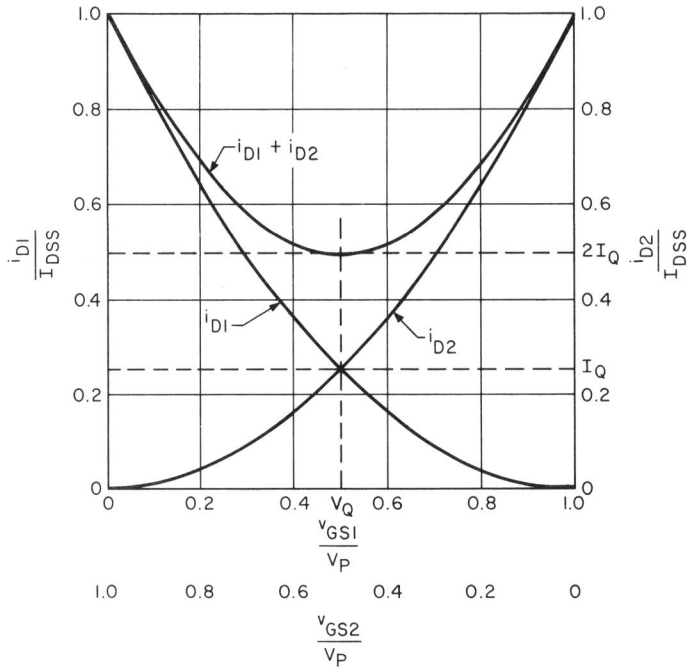

Fig. 4.2. Characteristics of FET squaring circuit.

parabola, with the zero or lowest level of the curve being twice the value of the bias level of either transistor. An input sine wave is shown superimposed on the parabolic transfer curve; the output is proportional to $\sin^2 \theta$, where the constant of proportionality is

$$K = \frac{2I_{DSS}}{V_P^2}$$

Since

$$\sin^2 \theta = \frac{1 - \cos 2\theta}{2}$$

the output is twice the frequency of the input.

A circuit to implement these squaring properties at audio frequencies is shown in Fig. 4.3. The first stage is a phase divider whose outputs drive the squaring FET's Q_2 and Q_3. The output of the squarer is coupled through a capacitor to a meter rectifier circuit. The meter reading is proportional to the square of the amplitude of the input voltage. The graph inset in Fig. 4.3 illustrates the squaring property of the circuit. The output voltage drops to 25 per cent of its full scale reading when the input is decreased by 6 db (2 to 1). The photograph in Fig. 4.4 was taken of the waveform at the squarer output, i.e., at the drains of Q_2 and Q_3. Notice that the frequency is doubled, and that the output waveform is just about as clean a sine wave as the input, indicating the absence of harmonics other than the second. This fact is borne out by the data in Table 4.1, which tabulates the harmonic content of the output voltage. The data are given in terms of relative harmonic magnitudes in decibels, with the second harmonic taken as reference.

FET's in Nonlinear Circuits

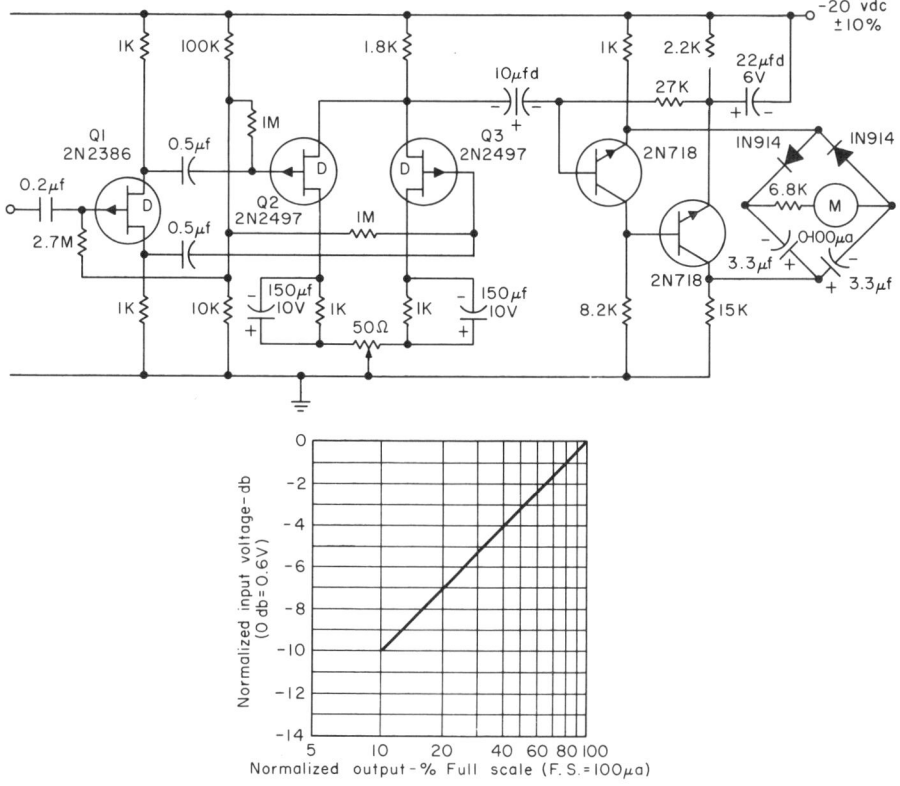

Fig. 4.3. FET squaring circuit and meter rectifier.

Photographs of the circuits response to two other input waveforms are shown in Figs. 4.5 and 4.6. If the triangular input waveform in Fig. 4.5 can be considered as repeating straight-line segments, the output is a series of repeating parabolas, one for each input straight-line segment. If the Fourier expansion of a perfect square wave is itself squared, all harmonics fall out leaving only a d-c term. Thus, the output in Fig. 4.6 is only the base-line trace shifted from the reference or qui-

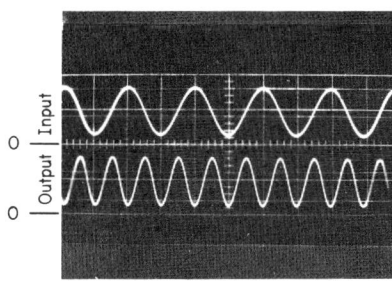

Fig. 4.4. Input and output waveforms.

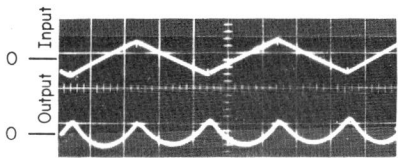

Fig. 4.5. Square of triangular waveform.

84 Field-effect Transistors

Table 4.1. Harmonic Rejection Properties of FET Squarer

Frequency	Harmonic output, db				
	1st	2nd	3rd	4th	5th
50 cps	−40	0	−40	−50	−54
1 kc	−46	0	−39	−54	−82
10 kc	−42	0	−39	−55	−82

escent drain voltage at the junction of Q_2 and Q_3 by a steady d-c voltage, which happens to be the square of the amplitude of both the positive and negative excursions of the input waveform.

The frequency response of the circuit in Fig. 4.3 to input signals is given by the graph in Fig. 4.7.

An important application of the squaring circuit in analog computation is in analog multipliers based on the "quarter-squares" principle. This method of multiplication is based on the identity

$$V_A V_B = \frac{1}{4}[(V_A + V_B)^2 - (V_A - V_B)^2] \quad (4.6)$$

How this operation would be performed by an electronic network is illustrated by the block diagram in Fig. 4.8. The availability of matched FET pairs in integrated-circuit packages suggests that the system of Fig. 4.8 might be built entirely of integrated-circuit components.

4.2 FET'S IN SWITCHING CIRCUITS

The important parameters of a switch are the ON resistance, the OFF resistance, and the opening and closing times. When the FET is used as a switch, the contact terminals are the drain and the source, and the gate is the control terminal. The analysis of electron-device switching circuits is usually broken up into three parts: the circuit and device combination is analyzed separately in the ON state, in the OFF state, and in transition from one state to the other.

Equivalent circuits for analyzing FET operation are shown in Fig. 4.9. The resistive elements determine the static ON and OFF behavior, and the gate capacitances are needed to determine speed of response during switching. The capacitances C'_{DG} and C'_{SG} in the OFF equivalent circuit are measured under higher reverse-bias conditions; consequently, they are somewhat smaller than C_{DG} and

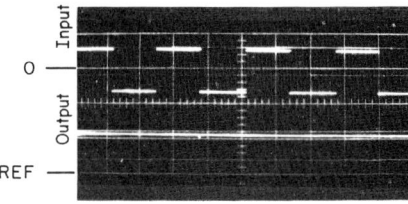

Fig. 4.6. Square of a square wave.

FET's in Nonlinear Circuits 85

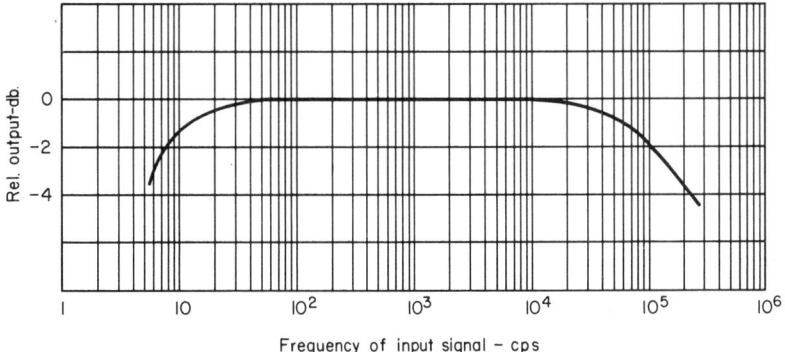

Fig. 4.7. Voltage squared-circuit frequency response.

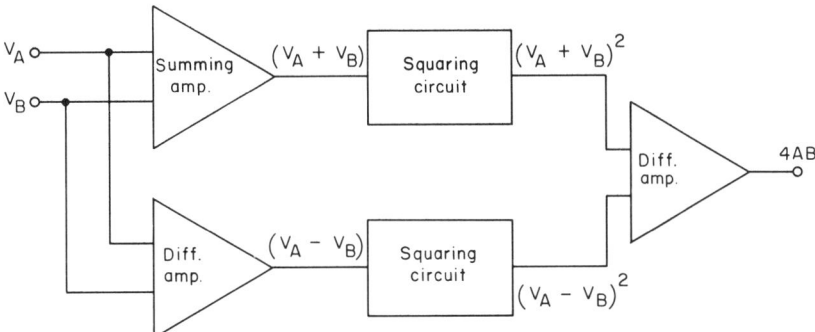

Fig. 4.8. Block diagram of quarter-squares multiplier.

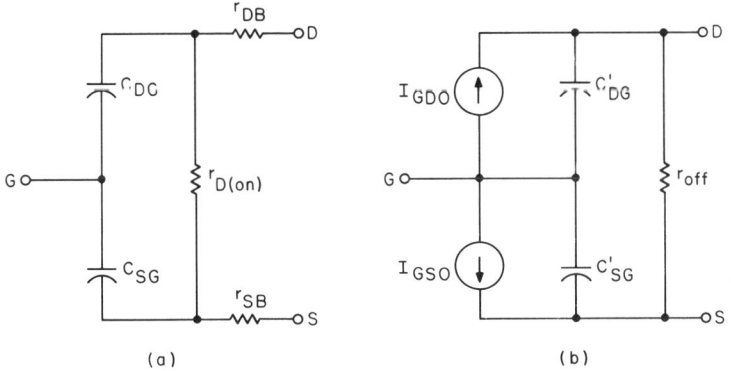

Fig. 4.9. Equivalent circuits of an FET switch: (a) switch ON; (b) switch OFF.

86　Field-effect Transistors

C_{SG}. As a matter of fact, the capacitances are nonlinear functions of the gate-to-channel voltage. They vary inversely as the cubed root of the applied voltage in diffused PN junctions, but a consideration of the OFF and ON end-point values alone yields sufficient accuracy in most cases. The minimum ON resistance in Fig. 4.9a is the sum of the drain and source bulk resistances and the reciprocal of transconductance when the gate is forward-biased by the contact potential ϕ. Of course, we have already seen that this is impossible to do, but it is possible to forward-bias the gate-to-channel diode to within 80 per cent of ϕ with only a few microamperes of forward gate current. Thus, in the FET, ON resistance is approximately

$$r_{on} = r_{DB} + r_{SB} + \frac{V_P}{2I_{DSS}}\left(\frac{V_P}{V_P + \phi}\right) \tag{4.7}$$

Inserting values of the above parameters for a hypical FET, the 2N2497 ($V_P = 2$ volts, $I_{DSS} = 2$ ma, $r_{DB} = r_{SB} = 75$ ohms) yields a typical value of r_{on} of approximately 550 ohms. This value of r_{on} is only valid for very small currents, owing to the nonlinear nature of the output characteristics. A simple worst-case approximation for $r_{D(on)}$ can be used when the ON currents are a significant percentage of I_{DSS}. Figure 4.10 shows voltage saturation curves. The line intersecting the origin and having the slope I_{DSS}/V_P will always yield a higher voltage than the true saturation curve at any current lower than I_{DSS}, and is a pretty good approximation at currents greater than I_{DSS}. Therefore, a useful approximation to the ON resistance is

$$r_{on} \cong \frac{V_P}{I_{DSS}} \tag{4.8}$$

The OFF resistance of most FET's is in the order of 10^9 ohms and can generally be considered an open circuit.

Multivibrators. The bistable multivibrator or flip-flop is a basic element of digital computing systems. Modern computers require flip-flops that run at ever higher speeds. As we pointed out previously, the FET is not yet equipped for this electronic derby, i.e., a speed competition with bipolar transistors. In some slow-speed applications, however, FET's may offer the advantage of being somewhat radiation hardened.

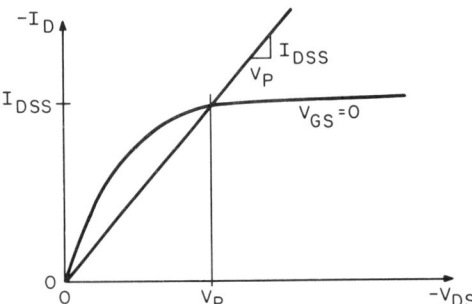

Fig. 4.10. Approximate saturation line.

Figure 4.11 shows a basic FET flip-flop circuit. Since speed is not a primary consideration, we will consider only the stable states.

ON equation: If Q_1 is ON and Q_2 OFF, then the saturation current that flows in the drain of Q_1 is approximately

$$I_{DS} = -\frac{V_{DD}}{R_D + \dfrac{V_P}{I_{DSS}}} \quad (4.9)$$

To ensure that the drain-to-source voltage will be less than V_P, the gate voltage ought to be zero volts or less, or

$$(V_{DD} + V_{GG})\frac{R_G}{R_G + R_F + R_D} \leq 0 \quad (4.10)$$

OFF equation: To ensure that Q_2 will remain OFF, the gate voltage must not be allowed to get within its "gate base"; that is, the gate-to-source voltage of Q_2 must be greater than V_P. This condition will be satisfied if

$$\left(I_{DS}\frac{V_P}{I_{DSS}} + V_{GG}\right)\left(\frac{R_G}{R_F + R_G}\right) \geq V_{GG} - V_P \quad (4.11)$$

Equations (4.11) and (4.10) must be satisfied simultaneously for a usable design; the worst-case values of all the terms in both equations must be considered, including resistor tolerances.

A circuit designed from these considerations is shown in Fig. 4.12. It is connected as a scale-of-two frequency divider (binary) complete with steering diodes, speed-up capacitors, and coupling capacitors.

FET monostable multivibrators are most useful as timers because the high input impedance allows the use of relatively modest-sized capacitors to obtain long time delays. A monostable circuit for use as a timer is shown in Fig. 4.13. Q_2 and Q_3 are a compound connection essentially similar to the transistor Darlington pair. This combination and Q_1 form the monostable multivibrator. In the stable state,

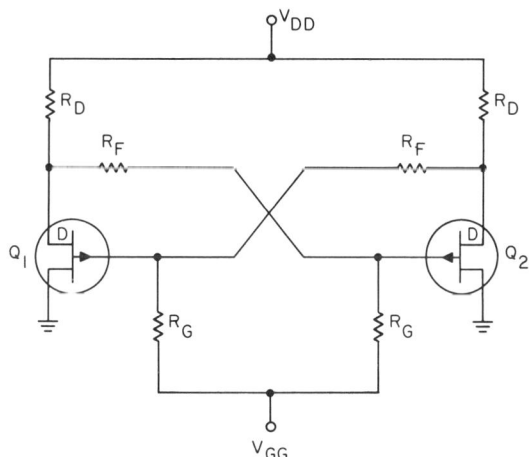

Fig. 4.11. Basic FET flip-flop.

88 Field-effect Transistors

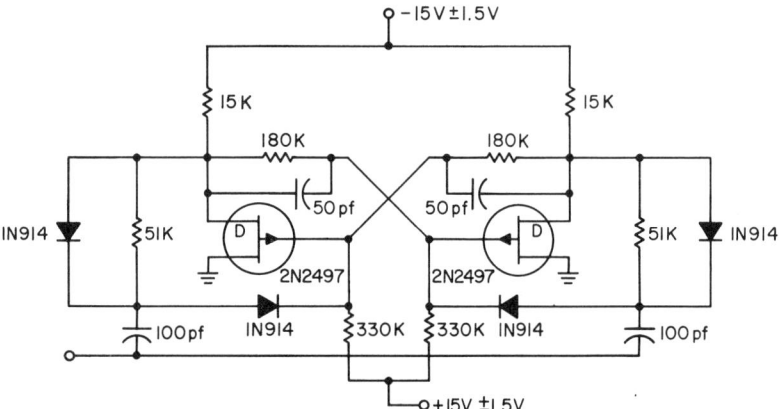

Fig. 4.12. An FET flip-flop.

Q_2 is ON and Q_3 is saturated, holding Q_1 OFF through the 15-K and 10-K resistive divider. Diode D_1 provides some reverse bias for the base-emitter diode of Q_1. The static voltage across C_1 is equal to the magnitude of the 8-volt negative supply minus the forward drops across three diodes: the forward voltage across D_1, the base-emitter forward voltage of Q_3 (V_{BEF}), and the forward gate-to-source voltage of Q_2 (V_{GSF}). When the push-button switch S_1 is momentarily closed, Q_3 turns OFF turning Q_1 ON almost instantaneously. The collector voltage at Q_1 changes by

$$\Delta V_{C1} = V_{DD} - V_{CES} - V_{D1} \tag{4.12}$$

where V_{CES} is the saturation voltage of Q_1. Since the voltage across C_1 cannot

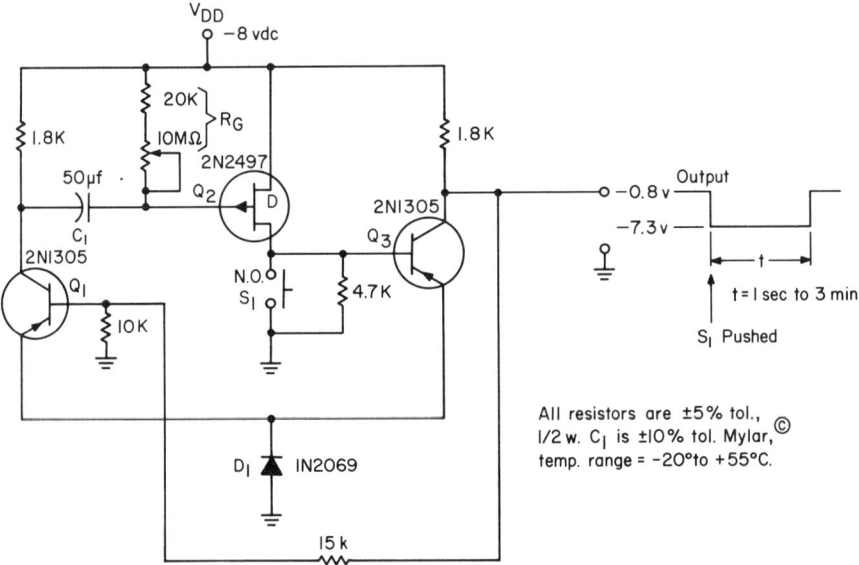

Fig. 4.13. Field-effect timer circuit.

change instantaneously, the voltage at the gate of Q_2 must change by the same amount. If this is greater than the pinch-off voltage of Q_2, Q_2 is turned OFF holding OFF Q_3 which in turn holds Q_1 ON. This state of affairs is maintained until C_1 discharges enough to allow Q_2 to conduct sufficiently to make the overall loop gain of the circuit exceed one, at which time the circuit rapidly returns to its stable state, waiting for S_1 to be pushed again.

Since C_1 must discharge through the variable gate resistor R_G, $C_1 R_G$ is the time constant of the circuit. Figure 4.14 is a voltage-time plot of voltage at the gate of Q_2; in the steady state, the gate rests at $-(V_{D1} + V_{BEF} + V_{GSF})$. At the instant of switching it jumps to $+(V_{DD} - V_{CES} - 2V_{D1} - V_{BEF} - V_{GSF})$. As indicated by the dotted continuation of the discharge curve, the gate would like to end up at $-V_{DD}$, but when the voltage drops to about V_P, switching occurs and the gate returns to its stable state. The delay time of the circuit is

$$T_D = R_G C_1 \ln \left[\frac{2V_{DD} - V_{CES} - 2V_1 - V_{BEF} - V_{GSF}}{V_{DD} - V_{CES} - 2V_1 - V_{BEF} - V_{GSF} - V_P} \right] \quad (4.13)$$

As indicated in Fig. 4.13, delays of three minutes are obtainable with this monostable circuit.

4.3 FET CHOPPERS AND COMMUTATORS

Choppers are used to convert d-c signals to a-c signals in amplifiers where significant thermal drifts cannot be tolerated. In a d-c amplifier, drift originating in any stage is transmitted to and amplified by all succeeding stages, and drift errors accumulate. An amplifier which responds only to a-c signals is not subject to this cumulative drift error. Consequently, in many d-c amplifiers, the input signal is changed into an a-c signal to be amplified and then rectified to give d-c output.

Many circuit configurations can be used for these amplifiers, but common to all is the need for a fast-acting switch placed in the signal path. This switch should not introduce any signal of its own, and should pass and block the intended signal without introducing any significant distortion.

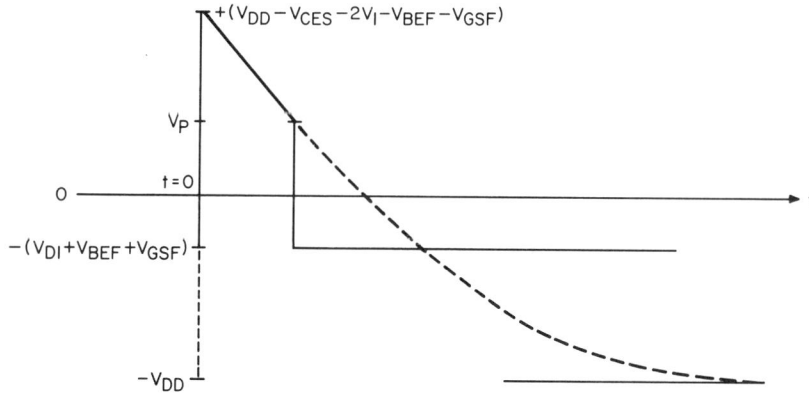

Fig. 4.14. Voltage-time diagram of gate voltage.

The same switch requirements are found for commutators in multiplex systems. Here the problem is to sample several signal inputs in rapid succession (time sharing) and to transmit these samples through a single common output point.

One of the outstanding features of an FET switch is its unmeasurably low offset or contact voltage when the switch is ON. There are (supposedly) no PN junctions in the current path between the drain and the source. The current path has a resistance whose value is given by Eq. (4.7); it usually runs between 100 and 5,000 ohms. This ON resistance is quite high compared to the ON resistance of bipolar transistors used as choppers—5 to 50 ohms. The extremely low offset voltage is not bought cheaply.

A P-channel FET is shown connected as a shunt and a series chopper in Figs. 4.15a and b respectively. In each case, the termination resistor R_{in} represents the input resistance of the a-c chopper amplifier.

The effect of the FET's high ON resistance can be made negligible by making R_{in} large compared to $r_{D(on)}$ in the series chopper circuit. In the shunt chopper, R_S should also be large compared to $r_{D(on)}$ to minimize error. This would not be necessary if the chopper operated at a constant temperature, since $r_{D(on)}$ causes an attenuation of the signal which is constant regardless of level, but $r_{D(on)}$ will not necessarily track with R_{in} or R_S over any temperature range. This effectively causes a change in gain with temperature.

In addition to high ON resistance, another problem in FET choppers is the overshoot or "spikes" that occur in the amplifier input at the chopper transition times. This is the result of the gate capacitance coupling into the load the fast changes in the gate drive voltage. The way to reduce gate capacitance is to reduce the dimensions of the channel. Reducing W in Fig. 1.15 also causes the ON resistance to increase, and this is undesirable; but by reducing L, both the capacitance and the ON resistance can be reduced simultaneously. How much L can be reduced depends entirely on the photomasking process by which the FET is made. Nothing is gained by reducing the channel height, since this increases the pinch-off voltage, increasing chopper drive requirements. Any reduction of transient current resulting from lowering the gate capacitance is more than offset by the increased V_P. This can be verified by inspecting Eqs. (1.8) and (1.9).

An FET with a foreshortened channel (in the x direction) will tend to be governed by velocity-limited current flow rather than the pinch-off mechanism; i.e.,

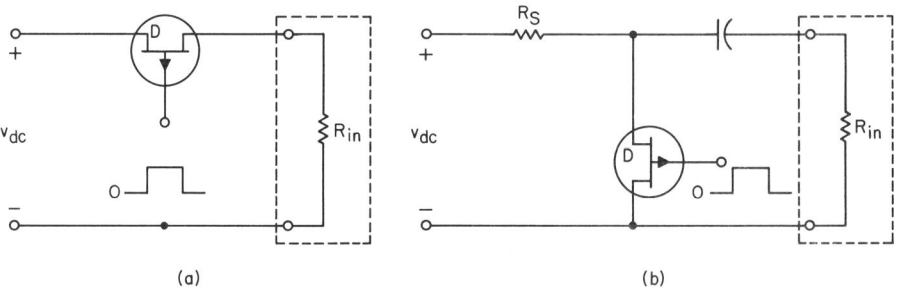

Fig. 4.15. (a) Series chopper; (b) shunt chopper.

FET's in Nonlinear Circuits   91

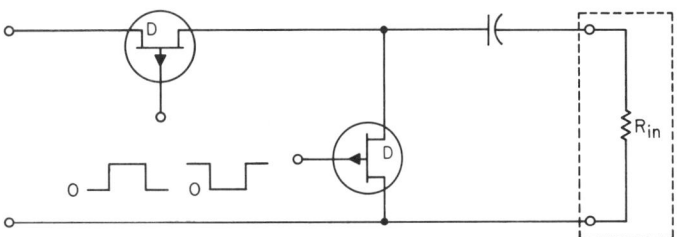

Fig. 4.16. Series shunt chopper.

the channel will not narrow down to insignificant dimensions when carrier velocity limiting occurs. Intuitively, one would expect this to make the output impedance relatively low in the pinch-off region, but obviously no problem will arise here since it is desirable for a chopper to be either completely ON or completely OFF.

The lowest level voltage a chopper is capable of measuring is determined ultimately by the noise in the series resistance "seen" by the chopper a-c amplifier input. This resistance is determined by the internal resistance of the voltage source being measured, and either R_S or the chopper ON resistance, depending on whether a shunt or series chopper is being used. Obviously, a shunt chopper is undesirable for very low levels; the series-shunt arrangement in Fig. 4.16 will work better than a simple series chopper. The two FET's are switched alternately, the shunt switch keeping the noise level down when the series switch is open. Although the two gates are driven out of phase and the spikes should tend to cancel, the FET capacitances are nonlinear; one FET is being switched from low to high capacitance, while the other is being switched from high to low.

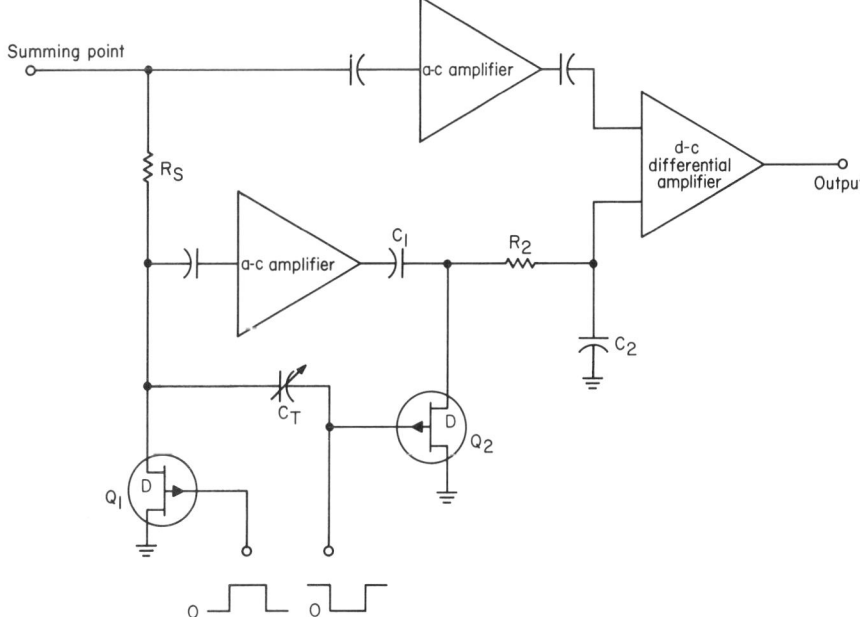

Fig. 4.17. Chopper stabilized d-c amplifier.

The capacitances will always be unbalanced. This means the spikes will be large; so this type of chopper will generally have to be operated at a relatively slow speed, say 400 cycles or less.

Figure 4.17 shows how an FET chopper would be used in a chopper-stabilized d-c amplifier. FET Q_2 is the demodulator; when it is ON_1, C_1 charges up rapidly to the signal voltage. When Q_2 turns OFF, Q_1 turns ON, referencing the input to zero and the voltage across C_1 is applied to the low-pass filter R_z and C_z. C_T is a small trimmer condenser that helps minimize the spikes at the input of the a-c amplifier.

Throughout this discussion, we have limited ourselves to P-channel FET's. Because the mobility of electrons in silicon is three times that of holes, the N-channel FET's when generally available, will make better choppers. The ON resistance will be one-third that of P-channel FET's, all dimensions being equal.

5

Blue Skies Dept.: The Power FET

The first junction transistors developed were small-signal devices. Power transistors worthy of the name did not begin to appear until the middle fifties, at least three years after transistors became commercially available. In the present development of field-effect transistors, most of those available thus far can be classified as small-signal, but at least one manufacturer (Amelco) has a medium-power FET on the market ($V_{DS} = 200$ volts, $I_{D(on)} = 25$ ma). As sure as the power transistor was developed, we can expect to see higher-power FET's soon.

A critical appraisal of the FET's development might raise the question of the utility of such apparent "development for development's sake." Or, "why go to the expense of development when bipolar transistors are available and in such an advanced state of development?" The validity of this criticism can be tested by pointing out the anticipated advantages and disadvantages of power FET's *vis-à-vis* power transistors. Of course, power FET's already have a big disadvantage in this comparison—there are none. Their behavior at high power levels will have to be inferred from their low-level performance. Such a procedure is subject to error within certain tolerance limits.

5.1 THE POWER FET AND POWER TRANSISTOR COMPARED

Until now, bipolar power transistors have shown certain deficiencies:

1. Lack of frequency response
2. Secondary breakdown
3. Thermal instability
4. Complexity of driver circuits required

Transistors rugged enough to stand continuous operation at high power levels without a serious threat of secondary breakdown generally have a rather modest gain-bandwidth product (f_T) of from 100 kc to 1 mc. On the other hand, power transistors that possess f_T in the neighborhood of 10 mc or higher are subject to secondary breakdown at much lower power levels than the thermal properties of the transistors might imply. While there is no evidence one way or the other concerning the FET and secondary breakdown, there is good reason to expect that

94   Field-effect Transistors

a power FET would not be subject to this phenomenon. Its mechanism involves neither the injection of current carriers across a PN junction, nor the collection of carriers in an intense electric field. Presence of one or both of these phenomena is required to support existing theories on secondary breakdown.

The frequency response of a power transistor in a common-emitter stage is determined by its "beta cutoff" frequency which, even for diffused transistors, is in the order of 200 kc or less. The cutoff frequency of the common-source transconductance of present field-effect transistors is above 20 mc. The input capacitance and equivalent input resistance are the limiting factors in determining frequency response. By driving the power FET with a low-output-impedance driver, such as an emitter-follower or source-follower, the frequency response can be made to compare favorably with common-emitter diffused power transistors.

5.2   THERMAL STABILITY AND DRIVER REQUIREMENTS

Thermal runaway must be guarded against when using power transistors. Because of a transistor's thermal feedback, power dissipated in the transistor raises the junction temperature, which in turn increases not only the leakage current but also the static and dynamic transconductances, causing the dissipated power to increase further. This regenerative action continues until something unpleasant happens. Transistor stages stabilized against thermal runaway have some external emitter resistance and the external base resistance is very low, usually less than 10 ohms. This lowers efficiency and requires complicated, usually direct-coupled, driver circuits.

The FET's leakage current increases with temperature in the same manner as the I_{CBO} of a power transistor, and thus causes increased dissipation. But, unlike the transistor, both the static and dynamic transconductances of an FET decrease with increasing temperature. The power FET, therefore, should not be as prone to thermal runaway as is the power transistor. The bias could probably be stabilized with a relatively large gate resistor (10 K to 100 K) and no external source resistance. The driver stage need only be a large-signal voltage amplifier biased at a few milliamperes.

5.3   HYPOTHETICAL POWER FET

Output Power Considerations. The characteristics required of a power FET are determined by the output power required of a given circuit. The hypothetical device described here is one which provides 30-watts output from a push-pull output transformerless stage like Fig. 5.1. The graph of the output characteristic and load line in Fig. 5.2 shows which quantities should be specified for a given output power. From Fig. 5.2:

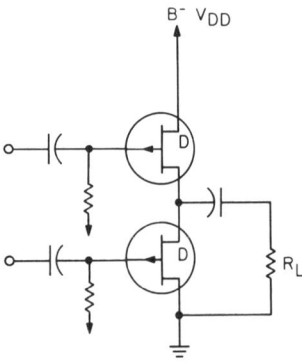

Fig. 5.1. A class B push-pull FET amplifier.

$$P_{o(max)} = \frac{V_{peak}I_{peak}}{2} = \frac{(V_{DD} - 2V_{sat})}{4} I_{DSS} \quad (5.1)$$

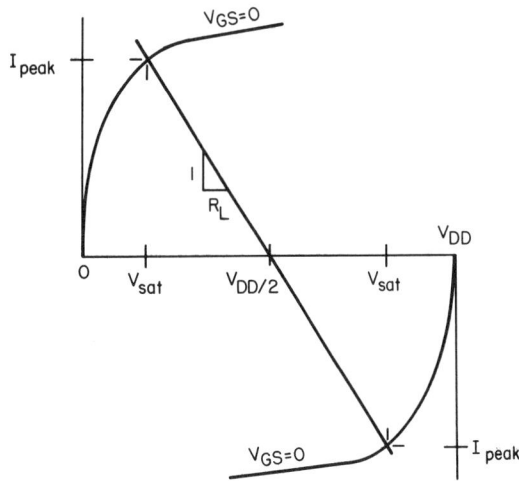

Fig. 5.2. Load line for push-pull amplifier.

Another consideration is efficiency; the maximum efficiency is

$$\eta_{max} = \frac{P_{o(max)}}{P_{in}} = \frac{\pi}{4}\left(1 - 2\frac{V_{sat}}{V_{DD}}\right) \tag{5.2}$$

If a 50-volt supply is used and at least 60 per cent efficiency is required of the circuit, then the upper limit on the saturation voltage is

$$V_{sat} \leq \frac{V_{DD}}{2}\left(1 - \frac{4}{\pi}\eta_{max}\right) = 6 \text{ volts} \tag{5.3}$$

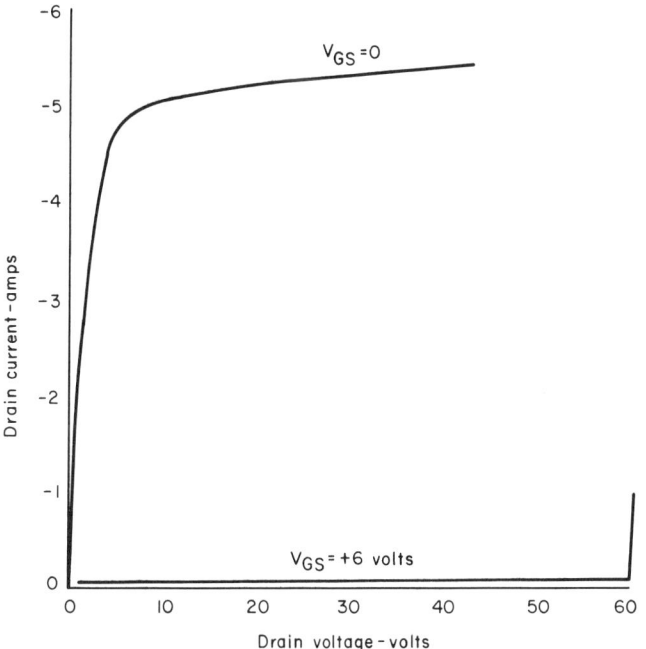

Fig. 5.3. P-channel power FET hypothetical characteristics.

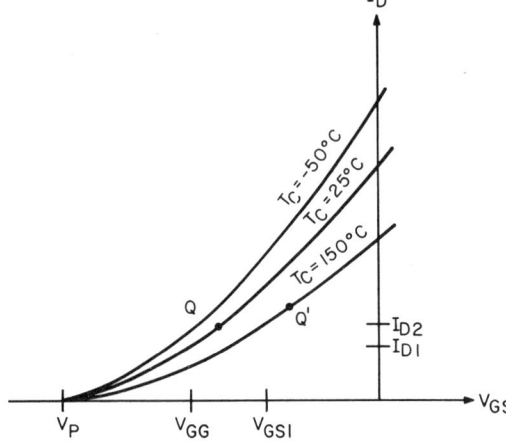

Fig. 5.4. Transfer curve variation with temperature for power FET with high V_P.

For 30-watts output, I_{DSS} must be at least

$$I_{DSS} \geq \frac{4P_{o(max)}}{V_{CC} - 2V_{sat}} = 3.16 \text{ amps} \tag{5.4}$$

This must be the value of I_{DSS} at the highest operating channel temperature. At this temperature, I_{DSS} falls to about 65 per cent of its room-temperature value. The room temperature I_{DSS} should be 4.9 amps. The supply voltage of 50 volts requires that the breakdown voltage BV_{GDO} be at least 50 volts plus twice the pinch-off voltage, or about 60 volts. These results are combined in the static output characteristics in Fig. 5.3.

Stabilizing the Bias of a Power FET. The behavior of I_{DSS} with temperature makes it possible to stabilize the quiescent drain current without using a source resistor. Consider Fig. 5.4, which is the typical transfer characteristics of a diffused FET, with channel temperature as a running parameter. If the FET is fixed-biased with a voltage V_{GG}, the 25°C operating point is Q, and the drain current is I_{D1}. To ensure that the drain current will not exceed an arbitrary I_{D2} at the highest operating channel temperature (in this example $T_C = 150°C$), it is necessary to make

$$R_G \leq \frac{V_{GG} - V_{GS1}}{I_{GSS(max)}} \tag{5.5}$$

Figure 5.5 is a graph of the quiescent drain current vs. temperature.

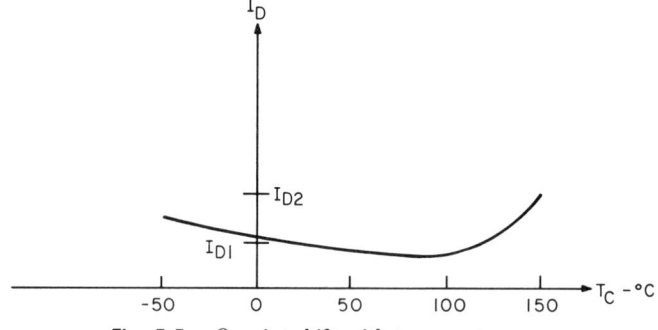

Fig. 5.5. Q-point shift with temperature.

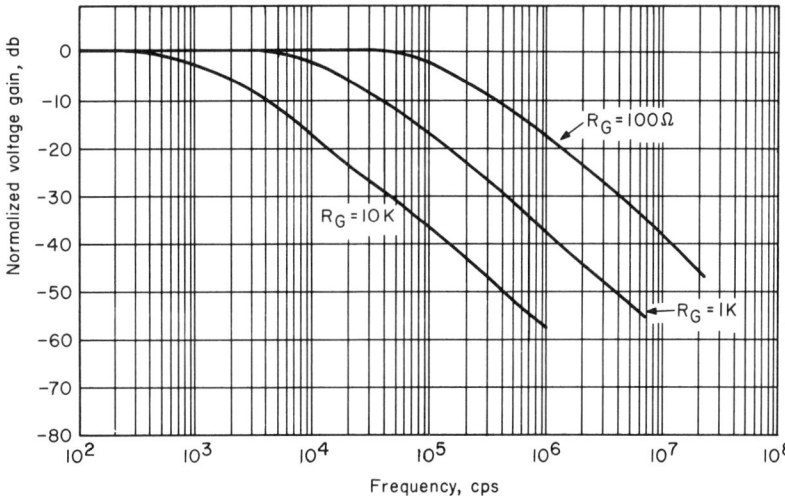

Fig. 5.6. Frequency response of hypothetical power FET.

Frequency Response. The frequency capability of the power FET can be roughly determined by taking the y parameters of a 2N2499 at $I_D = 10$ ma and multiplying them by 300 to simulate the y parameters of a 3-amp FET. This amounts to the assumption that to increase the current capability of an FET, it is only necessary to increase the channel cross section.

The graph of Fig. 5.6 is the normalized small-signal voltage gain at $I_D = 3$ amps, calculated from the two-generator equivalent circuit with an assumed load resistor of 8 ohms. The curves are plotted for three values of generator resistance. Values of generator resistance less than 1 kilohm can be obtained in practice by driving the power FET with an emitter-follower.

5.4 PRACTICAL AUDIO AMPLIFIER

Since no power FET's are available, a pair of medium-power devices were simulated by paralleling twenty 2N2386's (with I_{DSS} about 10 ma per device). The composite output characteristic of such a device is shown in Fig. 5.7. A pair of FET's with these characteristics should be capable of delivering about 600 mw into a 56-ohm load. Because the FET maintains its high input impedance at a few tenths of a volt forward bias, slightly more output power was obtained (about 750 mw) before clipping.

The circuit in Fig. 5.8 was used to test the capability of this device. Note the simplicity of the circuit. It can be described briefly as a voltage amplifier followed by a split-load phase inverter followed by the push-pull output stage. Emitter-followers are used to drive the output stage to improve frequency response; note that they are coupled to the output stage by nonelectrolytic capacitors. The performance data for this circuit are shown in Figs. 5.9 to 5.11.

Even though the power output of this circuit is less than 1 watt, it is easy to see that only the power devices are lacking to make this a 20- or 30-watt amplifier. Little, if any, change need be made in the rest of the circuit.

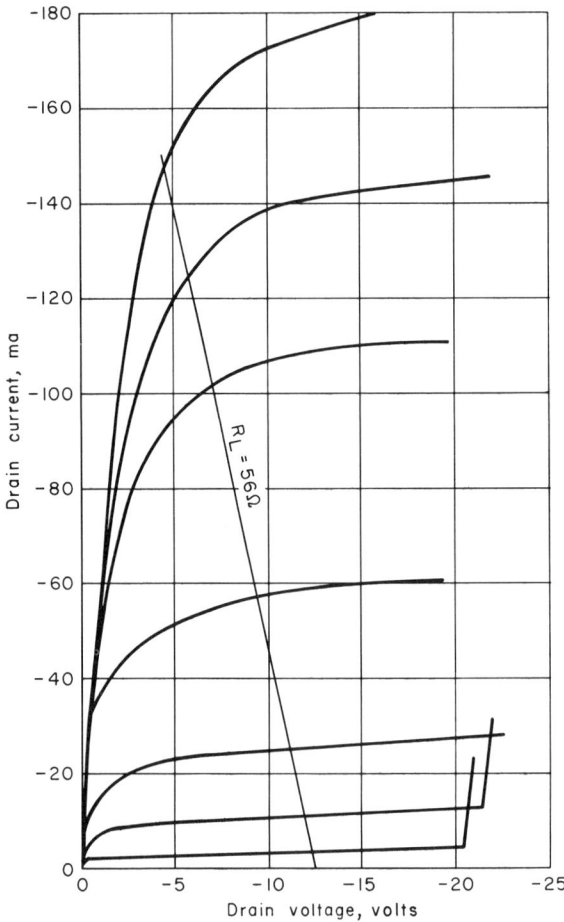

Fig. 5.7. Composite drain characteristics of twenty paralleled 2N2386's.

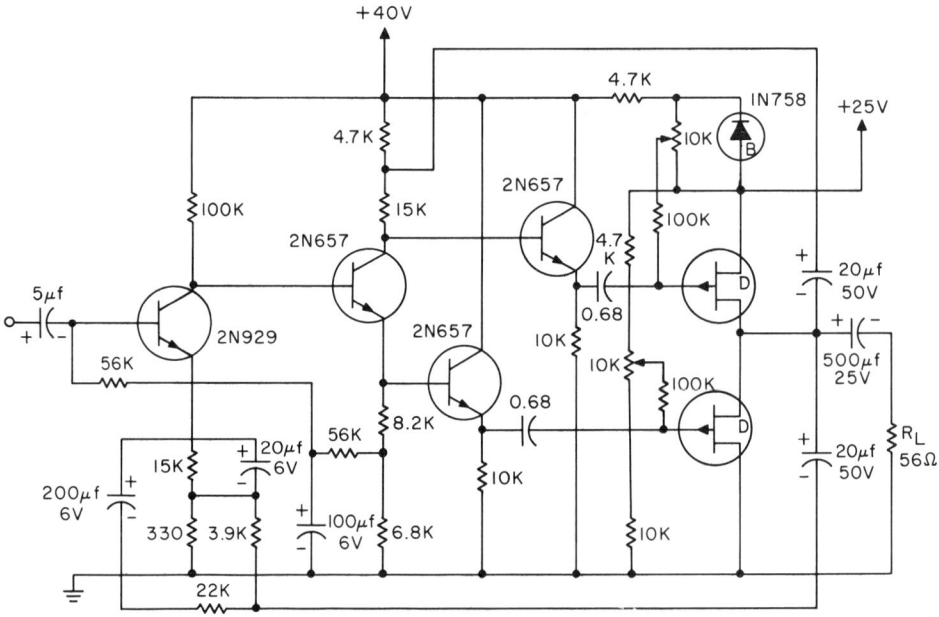

Fig. 5.8. Medium-power FET amplifier.

Blue Skies Dept.: The Power FET 99

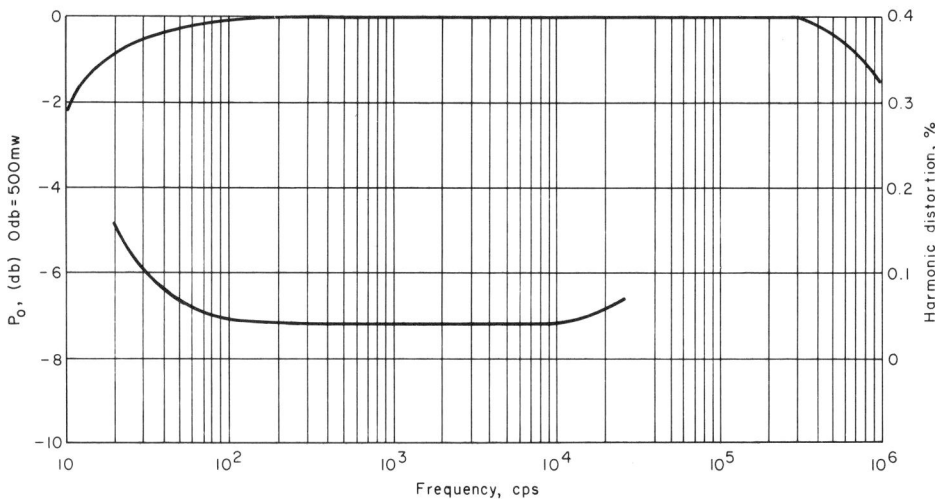

Fig. 5.9. Response curve.

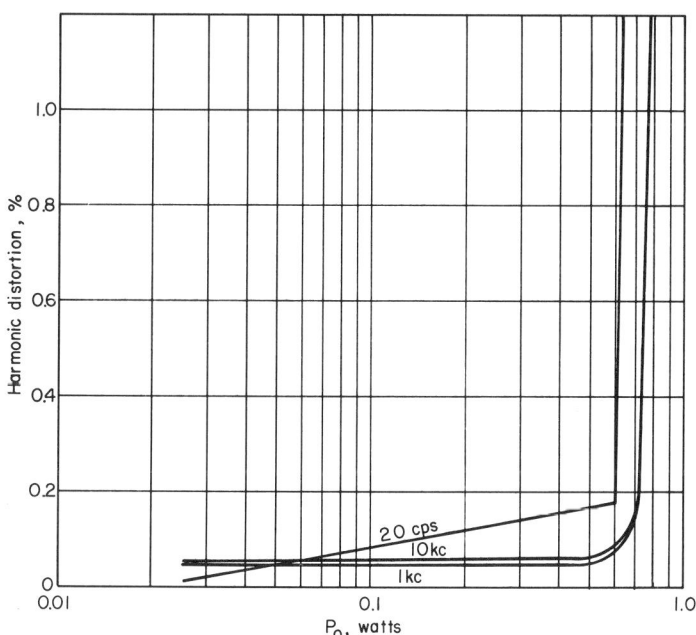

Fig. 5.10. Harmonic distortion vs. power out.

100 Field-effect Transistors

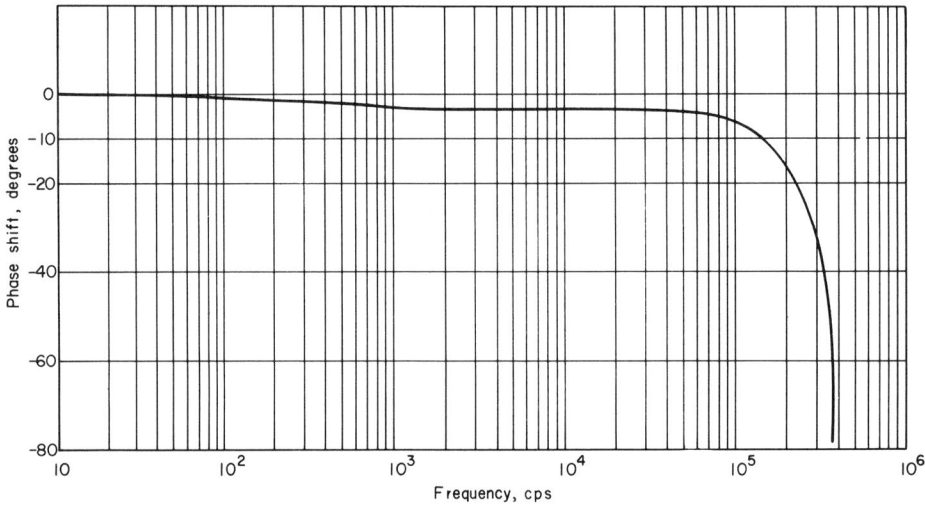

Fig. 5.11. Phase shift vs. frequency.

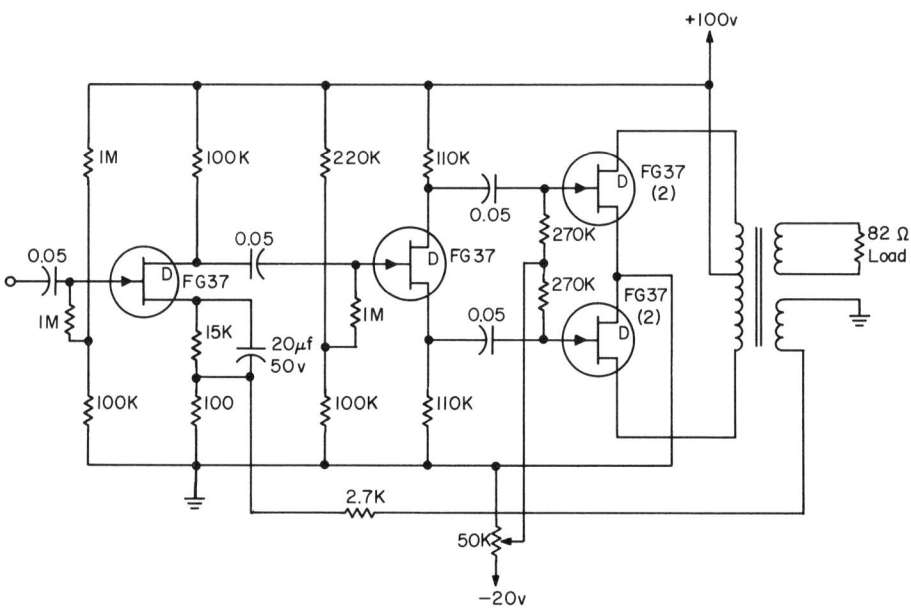

Fig. 5.12. 1.5-watt servo amplifier.

5.5 SERVO AMPLIFIER

Figure 5.12 shows a 1½-watt servo amplifier using Amelco medium power FET's. Note that the circuit has no driver transformer for the power stage, and that there is only one electrolytic capacitor (of a rather modest size) in the entire circuit. The amplifier has the following performance data:

Power gain	70 db
Voltage amplification	30 db
Input resistance	1 megohm
Maximum efficiency	56 per cent

6

Further Applications

The following is a compilation of some of the more interesting and useful circuits designed to date with FET's. Some practical circuits have been presented in the previous chapters, but in each case the purpose was to illustrate a particular development. It was felt that more than one illustration per idea would be distracting; so all the distractions have been saved for this chapter.

6.1 D-C AMPLIFIERS

Unity-gain Temperature-stable D-C Amplifier. In the amplifier in Fig. 6.1, each base of the 2N2641 dual transistor is driven by the source-follower FET's Q_1 and Q_2. Q_1 performs an impedance transformation while Q_2 closes the feedback loop and tends to cancel changes in parameters caused by temperature variations. The

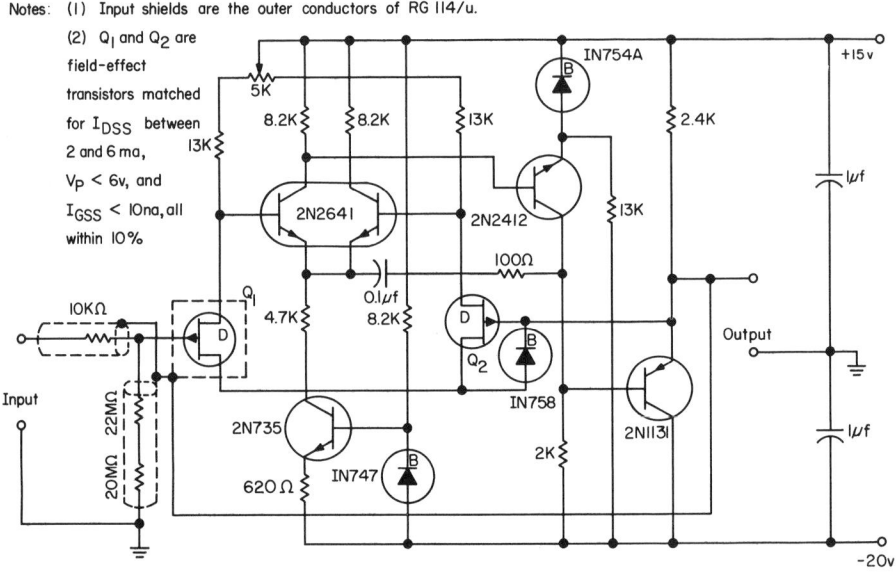

Fig. 6.1. Unity-gain temperature-stable d-c amplifier.

FET's are not, however, differentially connected, for the 1N758 breakdown diode connected to the output overrides any changes in FET current. The purpose of this breakdown diode is to force the voltage of the 2N735 current source, together with the drain voltage of the FET's, to follow the input voltage. This regulates the voltage across the gate-to-drain capacitance of Q_1, significantly decreasing the input capacitance and thereby extending the frequency range. To further reduce the input capacitance, the output is used to drive shielding around the entire input circuitry.

The output is taken from a 2N1131 PNP transistor used as an emitter-follower. At a frequency of 1 kc, the measured output impedance (with shorted input) is 0.35 ohm. For evaluation of the gain performance a 10-kc square-wave input signal with an amplitude of -5 volts was used. The difference between the output voltage and the input voltage was read on a Tektronix Type 581 scope with a type Z plug-in. Tracings of photographs taken over the temperature range of -55 to $85°C$ are shown in Fig. 6.2a–d for a generator resistance of 33 kilohms. The vertical sensitivity is 100 mv/division or 2 per cent error/division for a 5-volt signal. The zero-error line is the center line; the output was adjusted to zero using the 5-kilohms balance potentiometer. The left half of each tracing in Fig. 6.2 shows the error at the -5-volt level; the right half shows the error at the zero level. Because of the delay time in the amplifier, the initial error is nearly 5 volts, decreasing with time in a manner dictated by the transient response.

The d-c gain error may thus be defined as the average error over a definite period of time after the major transients have subsided. This error is less than 2 per cent over a 25-μsec period after turn-on, for a range of temperatures from somewhat above $25°C$ to below $-55°C$. At $85°C$ the d-c error is less than 3 per cent. Low-frequency a-c error decreased from 1.3 per cent at $85°C$ to 1.1 per cent at $-55°C$. Input impedance at very low frequencies is limited to 42 megohms by the two resistors connecting the gate of the input field-effect transistor to ground. These resistors can be removed if the generator impedance provides a d-c return to ground. Input impedance will then be greater than a 1,000 megohms at several cycles per second.

Figure 6.3 contains frequency response curves for generator resistances of 51 ohms, 33 kilohms, 100 kilohms, and 1 megohm. The best response is produced by the 100-kilohm generator. For that value, the response is down 3 db at a frequency of 2.55 mc.

Third-order Low-pass Active Filter. Amplifier circuits used in active filters must meet two requirements: (1) the gain must be very stable, and (2) the amplifier terminal impedances must not significantly affect the passive impedances in the branches of the network.

A specific example of an active filter is the third-order low-pass filter shown in Fig. 6.4. To satisfy the second requirement given above, the amplifier must have a very high input impedance, so that it does not load the R_3, C_3 branch of the circuit. The amplifier designed to satisfy both requirements is shown in Fig. 6.5. This is a more elaborate form of the bootstrapped source-follower of Fig. 3.18. The drain of the input FET drives a PNP transistor in cascode to reduce input capacitance. A 6.2-volt breakdown diode, the 1N751, supplies the bias voltage for

104 Field-effect Transistors

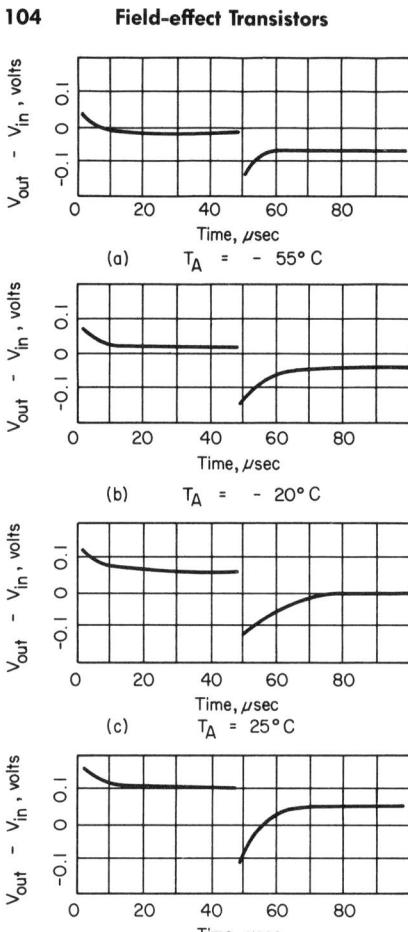

Fig. 6.2. Error in d-c amplifier.

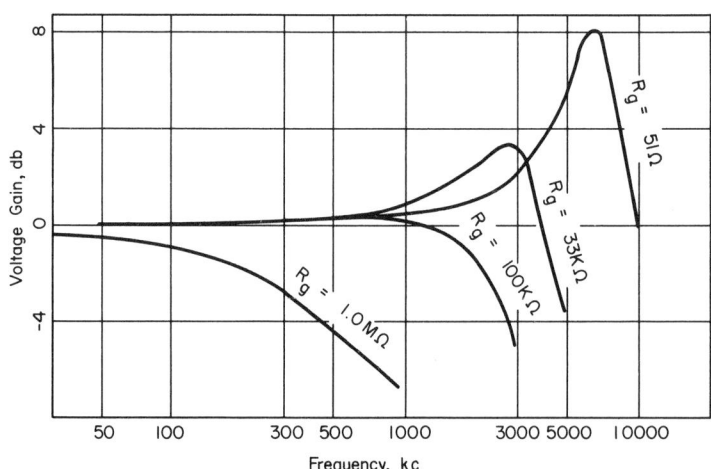

Fig. 6.3. Frequency response.

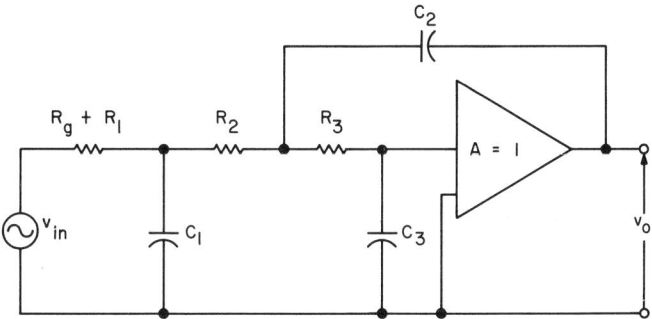

Fig. 6.4. Third-order low-pass active filter, general form.

the cascode pair. The source of the FET is driven in phase with the input through an emitter-follower, whereas in Fig. 3.18 it was driven directly by the collector of the NPN feedback transistor. The FET bias current is set by the resistance in the collector circuit of the 2N2411; the adjustment is made by potentiometer R_4 for zero d-c offset in the amplifier. The bias current is in the 200 to 400 μa range (depending on device selection), approximately the zero drift point of the 2N2497.

The transfer function of the filter is

$$T(s) = \frac{V_o(s)}{V_{in}(s)} = \frac{1}{S^3 + 1.75\,\omega_o S^2 + 2.15\,\omega_o^2 S + \omega_o^3} \quad (6.1)$$

where $\omega_o = 1.25$ radians/sec.

The normalized values of the passive elements are:

$$R_g + R_1 = 0.5 \quad\quad C_1 = 1.44$$
$$R_2 = 1.0 \quad\quad C_2 = 1.743$$
$$R_3 = 1.0 \quad\quad C_3 = 0.404$$

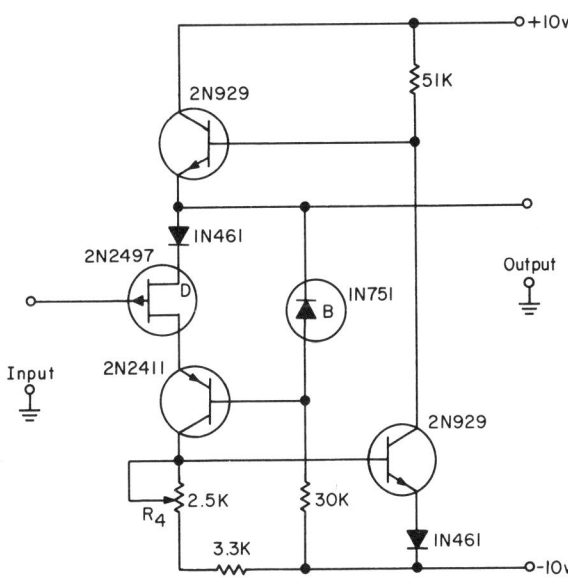

Fig. 6.5. Unity-gain amplifier for active filter.

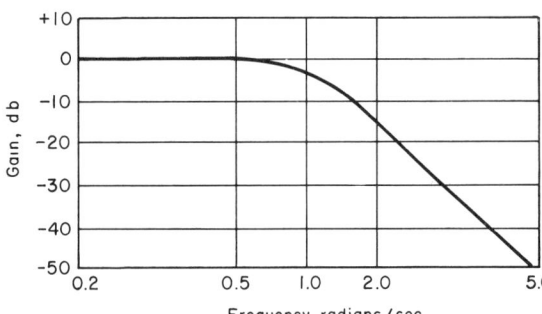

Fig. 6.6. Response curve.

Obviously, it is desirable to have R_1 large compared to the generator resistance R_g.

The normalized response curve is shown in Fig. 6.6. The impedance may be scaled by multiplying resistors by R and capacitors by $1/R$, where R is the new impedance level in ohms. The cutoff frequency may be scaled by multiplying capacitors by $\frac{1}{2}\pi f_c$, where f_c is the new cutoff frequency in cycles per second.

Example: Scaling the impedance to 100 kilohms gives

$$R_1 = 50 \text{ kilohms}$$
$$R_2 = R_3 = 100 \text{ kilohms}$$
$$C_1 = 14.4 \text{ μf}$$
$$C_2 = 17.43 \text{ μf}$$
$$C_3 = 4.074 \text{ μf}$$

and

$$\omega_c = 1 \text{ radian/sec}$$

Scaling the cutoff frequency to 100 cps gives

$$\omega_c = 628 \text{ radian/sec}$$
$$C_1 = 0.023 \text{ μf}$$
$$C_2 = 0.0277 \text{ μf}$$
$$C_3 = 0.00648 \text{ μf}$$

and resistances unchanged.

Scaling the cutoff frequency to 20 cps gives

$$\omega_c = 125.7 \text{ radians/sec}$$
$$C_1 = 0.115 \text{ μf}$$
$$C_2 = 0.1385 \text{ μf}$$
$$C_3 = 0.032 \text{ μf}$$

and resistances unchanged.

Operational Amplifier. An operational amplifier using discrete semiconductors with FET inputs is shown in Fig. 6.7.

The equivalent input at direct current is a 5-megohm minimum resistor, paralleled by a fixed current source equal to I_{GSS}. Q_1 and Q_2 should be reasonably matched in their transfer characteristics, to minimize the offset voltage when both

Further Applications 107

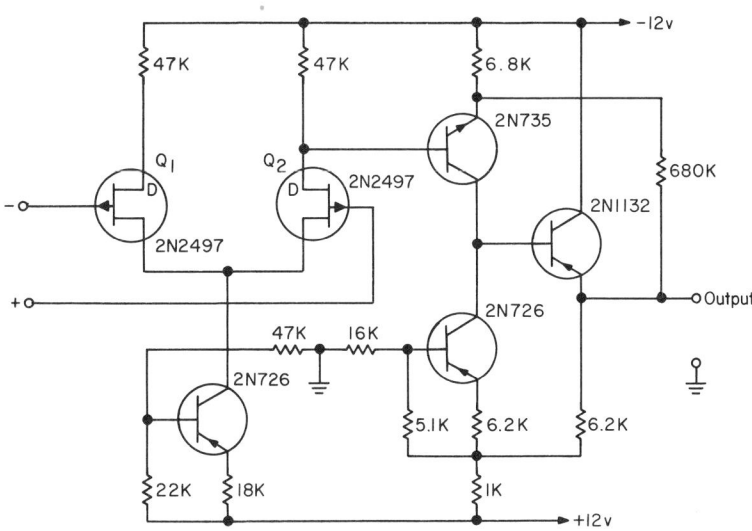

Fig. 6.7. Operational amplifier with FET inputs.

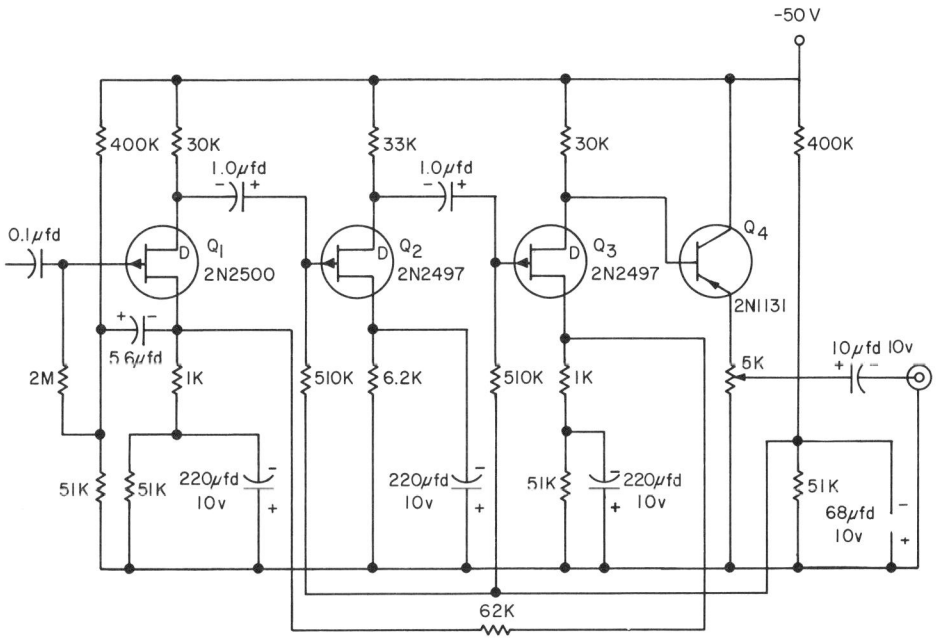

Fig. 6.8. 60-db low-noise amplifier.

108 Field-effect Transistors

inputs are grounded. The open-loop voltage gain at direct current is greater than 10^5; operated open-loop, this circuit makes an excellent voltage comparator having very high resolution.

6.2 A-C AMPLIFIER

The circuit in Fig. 6.8 is a three-stage all-FET voltage amplifier with an emitter-follower output. It can be used in applications requiring amplification of signals in the microvolt range. Some typical applications are: ultrasensitive preamplifiers for null detection apparatus, medical research electronic equipment, oscillographic and magnetic tape recorders; oscilloscopes, and all types of low-level transducers.

The maximum voltage gain is 60 db with a gain control provided. The frequency response is shown in Fig. 6.9. With a 100-kilohm generator resistance, the amplifier 3-db bandwidth is 1 cycle to 40 kc.

The input impedance is 100 megohms shunted by 15 pf. In most applications, this will be high enough to prevent any visible loading of circuits under test. The high input impedance is obtained through bootstrapping. For large variations in I_{DSS}, considerable increase in stability is achieved without loss in dynamic operating range by fixed-biasing the gates of Q_1, Q_2, and Q_3, and compensating for this by adding resistance to the sources of Q_1, Q_2, and Q_3.

The 2N2500 is used as the input stage because of its good low-noise characteristics at low frequencies. The noise curve in Fig. 6.10 shows the per-cycle noise as a function of frequency. Note that the noise output decreases with frequency. Since the total rms noise output is a composite of the per-cycle noise, the maximum signal-to-noise ratio will be achieved with the narrowest possible bandwidth.

The power supply requirement is 50 volts d-c at 6 ma.

6.3 ELECTRONIC VOLTMETERS

In the past, the abbreviation VTVM was synonymous with "electronic voltmeter." Now we are forced to discard the abbreviation. "Transistor VTVM" and "FET VTVM," while commonly used, hardly make sense.

Three circuits of varying levels of sophistication are presented here. First, a fairly simple (almost crude) FET voltmeter is shown in Fig. 6.11. The circuit

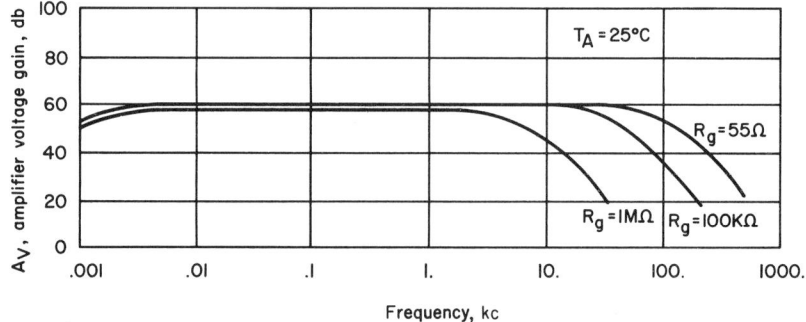

Fig. 6.9. Amplifier gain vs. frequency.

Further Applications 109

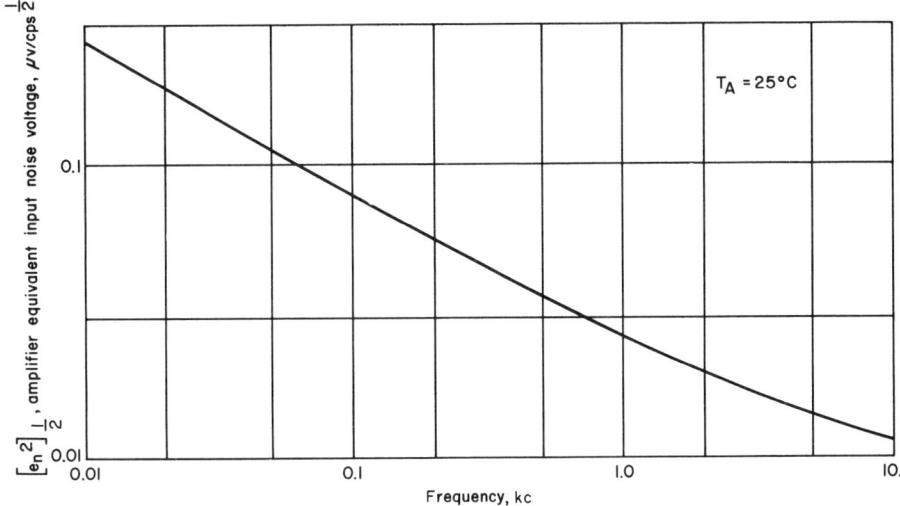

Fig. 6.10. Amplifier equivalent input noise voltage vs. frequency.

uses a single active device and is designed to indicate full scale at 1-volt input; its sensitivity is therefore 1 megohm/volt. The 2N2497 is biased at about 300-μa drain current, approximately the point of zero drift.

The voltmeter circuit in Fig. 6.12 is, in effect, two of the single-ended circuits of Fig. 6.11 connected "back-to-back" to form a differential voltmeter. Sensitivity of this circuit is the same as that of Fig. 6.11. The FET's should be matched for I_{DSS} and g_{max} to within 20 per cent.

A more complex and much more sensitive voltmeter is shown in Fig. 6.13. Q_1 and Q_2 should be matched to within 20 per cent on I_{DSS}, I_{GSS}, and g_{max}. This voltmeter consists of two circuits very much like the bootstrapped source-follower, differentially connected and fed by an active current source. An input of 50 mv produces full-scale deflection, making the sensitivity 20 megohms/volt.

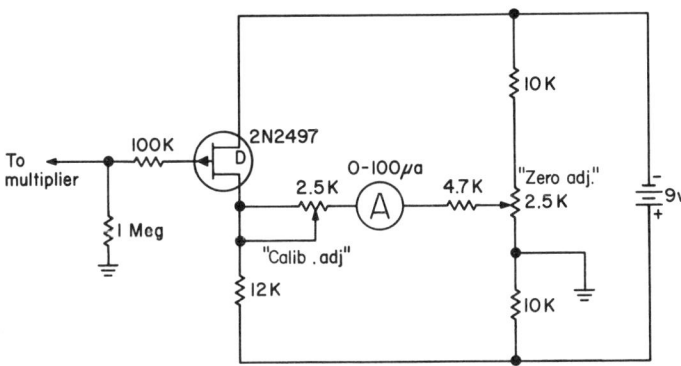

Fig. 6.11. Simple FET electronic voltmeter.

110 Field-effect Transistors

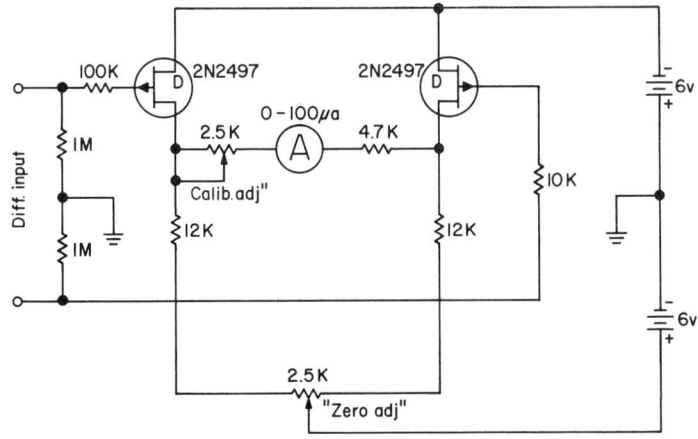

Fig. 6.12. Differential voltmeter.

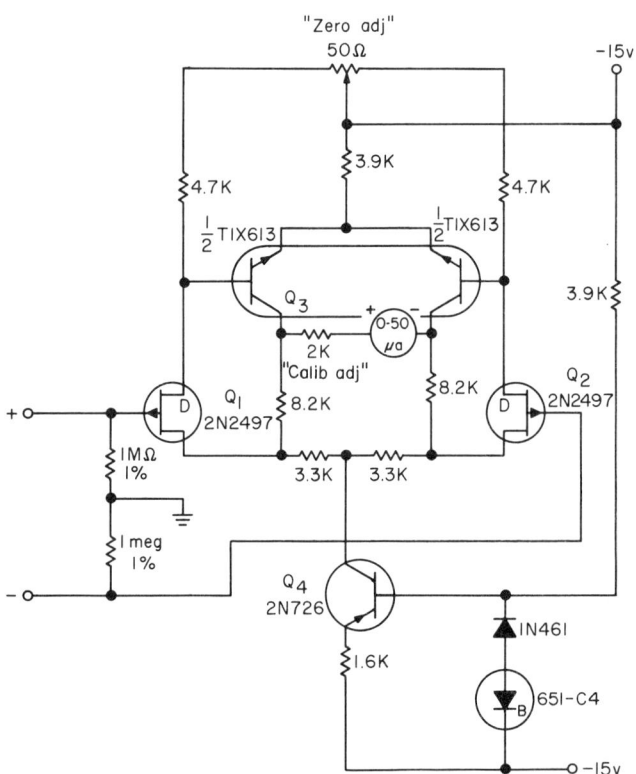

Fig. 6.13. D-c millivoltmeter, 0 to ±50 mv.

Further Applications 111

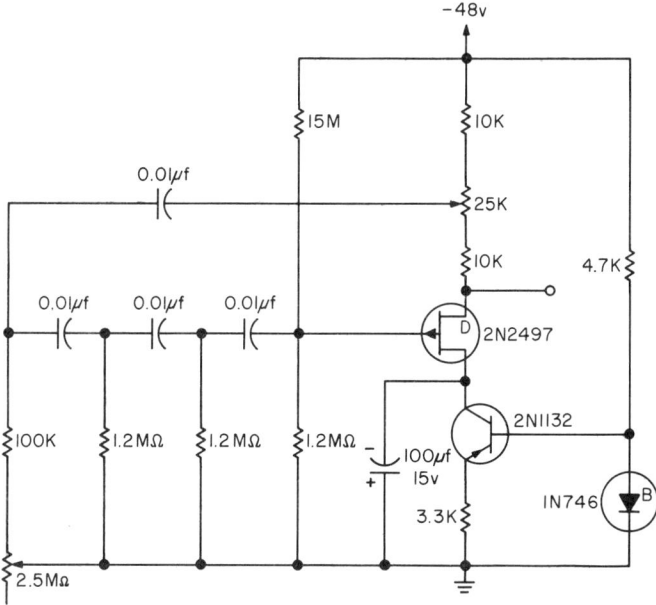

Fig. 6.14. Phase-shift oscillator.

6.4 OSCILLATORS

Phase-shift Oscillator. Figure 6.14 shows a four-mesh phase-shift oscillator, for use in applications where only limited frequency variation is necessary. The frequency of this oscillator can be varied several cycles around 10 cps, using the 2.5-megohm potentiometer. The attenuation of the four-mesh feedback network is 18.36. The frequency of oscillation is determined by

$$f_o \approx \sqrt{\frac{0.7}{2\pi RC}}$$

where R and C are the values of one mesh in the ladder feedback network. A three-mesh network would have an attenuation of 29 and a frequency of

$$f_o \approx \frac{1}{2\pi RC \sqrt{6}}$$

Wien-bridge Oscillator. A Wien-bridge oscillator is used where good amplitude stability is required for wide frequency variations. It takes the form of a two-stage RC-coupled class-A amplifier with two separate feedback loops between the output and input stage. One is a positive feedback loop that causes the oscillations, and the other is a negative feedback loop that stabilizes the amplitude of the oscillations. Figure 6.15a is a simplified schematic showing the general circuit configuration of a Wien-bridge oscillator. Figure 6.15b shows the feedback networks redrawn in the form of the bridge from which the oscillator derives its name. The positive feedback loop determines the frequency of oscillation; it is made up of R_1, R_2, C_1, and C_2:

112 Field-effect Transistors

$$f_o \cong \frac{1}{2\pi}\sqrt{R_1 R_2 C_1 C_2}$$

The attenuation at this frequency is 3.3. The other half of the bridge is a nonlinear negative feedback loop. B_1 and B_2 are tungsten filament lamps with positive temperature coefficients of resistance; they provide more feedback than can be obtained with a simple resistive divider. Since the lamps are operated at considerably less than rated voltage, they should have an extremely long life.

A practical Wien-bridge oscillator schematic complete with component values is given in Fig. 6.16. Three panel controls are necessary: a switch that selects resistors for a decade change in frequency; a variable capacitor for continuous frequency control within each decade; and an amplitude adjust control R_{14}. Frequency ranges are 20 to 200 cps, 200 cps to 2 kc, 2 to 20 kc, and 4 to 40 kc. The amplitude of the output signal can be adjusted from zero to approximately 3.5 volts rms with a 2-kilohm load. The circuit requires a 24-volt d-c 8-ma supply. The frequency stability on the low-frequency scale range is ±1.5 per cent; on all other ranges it is ±0.25 per cent. Harmonic distortion is less than 1 per cent on any scale range. Operating temperature range is 15 to 45°C.

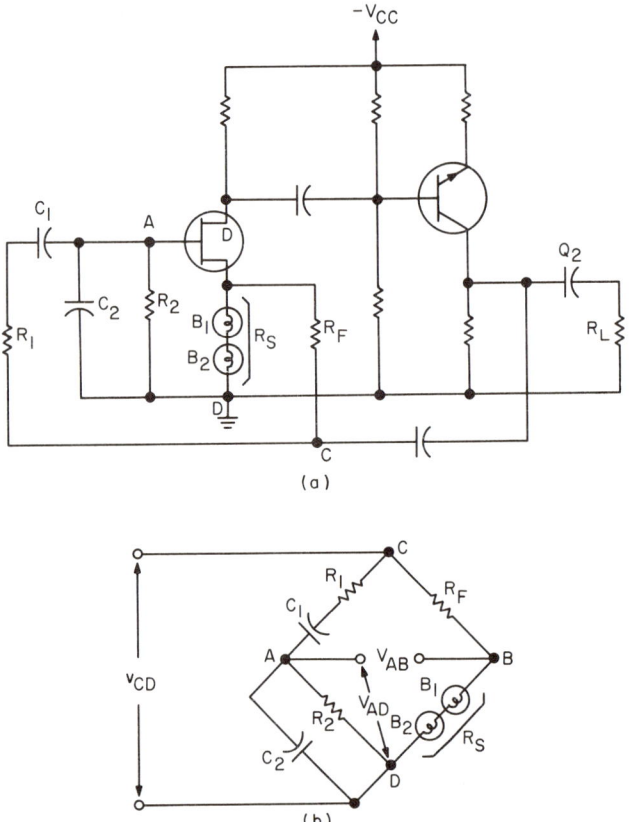

Fig. 6.15. Wien-bridge oscillator, simplified schematic: (a) oscillator; (b) feedback network.

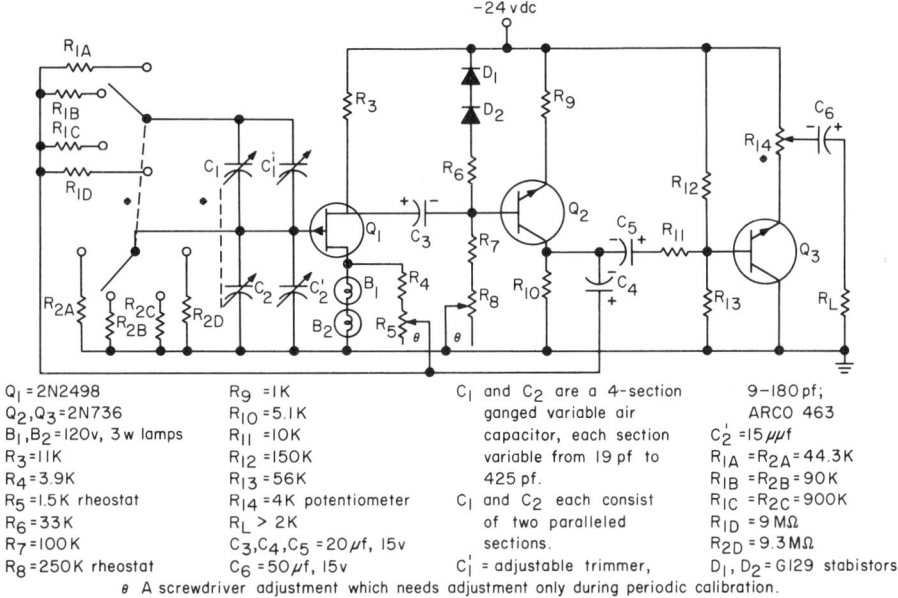

Q_1 = 2N2498	R_9 = 1K	C_1 and C_2 are a 4-section	9–180 pf;
Q_2, Q_3 = 2N736	R_{10} = 5.1K	ganged variable air	ARCO 463
B_1, B_2 = 120v, 3 w lamps	R_{11} = 10K	capacitor, each section	C'_2 = 15 $\mu\mu$f
R_3 = 11K	R_{12} = 150K	variable from 19 pf to	R_{1A} = R_{2A} = 44.3K
R_4 = 3.9K	R_{13} = 56K	425 pf.	R_{1B} = R_{2B} = 90K
R_5 = 1.5K rheostat	R_{14} = 4K potentiometer	C_1 and C_2 each consist	R_{1C} = R_{2C} = 900K
R_6 = 33K	R_L > 2K	of two paralleled	R_{1D} = 9 MΩ
R_7 = 100K	C_3, C_4, C_5 = 20 μf, 15v	sections.	R_{2D} = 9.3 MΩ
R_8 = 250K rheostat	C_6 = 50 μf, 15v	C'_1 = adjustable trimmer,	D_1, D_2 = G129 stabistors

θ A screwdriver adjustment which needs adjustment only during periodic calibration.
• A panel control.

Fig. 6.16. Variable-frequency Wien-bridge oscillator.

6.5 SAMPLE-HOLD CIRCUIT

A sample-hold circuit is a short-term analog memory. If an analog voltage is allowed to charge a capacitor through a mechanical switch contact, the capacitor will retain the analog information for a long time after the switch is opened. If the capacitor had no internal shunt leakage it would retain its charge indefinitely. But the capacitor voltage must be sensed or "read out" by some means, and the readout circuitry will draw some current, slowly discharging the capacitor and limiting the holding time. It is imperative then that the sensing amplifier draw as little current as possible. This obviously calls for an amplifier with an FET input stage.

A sample-hold circuit example is shown in Fig. 6.17. The input and sense amplifiers are identical. The input amplifier must have a low output impedance for rapid charging of the sampling capacitor C. It is also desirable that both amplifiers have unity gain with no d-c offset adjustment. Such an amplifier is shown in Fig. 6.18.

The input amplifier charges C through an FET switch, the TIX698. This is an FET designed especially for chopper applications. Its ON resistance $r_{D(on)}$ is very low, typically less than 100 ohms; this allows rapid charging of C. Another TIX698 is used to reset C after the sample time has ended. The sample-and-hold one-shots are conventional bipolar transistor monostable multivibrators. The delay time of the sample-one-shot must be long enough to allow C to charge to well within the desired accuracy of the circuit. Obviously, the delay time of the hold one-shot coincides with the hold time T_H. At room temperature, C will lose

114 Field-effect Transistors

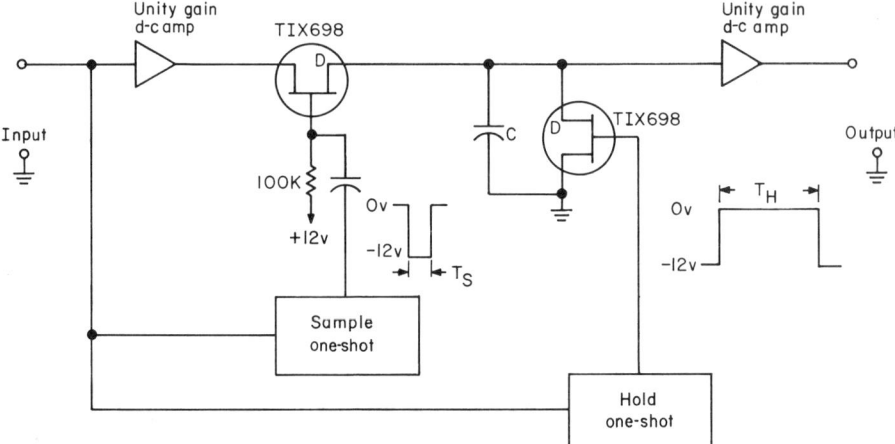

Fig. 6.17. Sample-hold circuit.

its charge at the rate of about 3 mv/μf-sec. The circuit will handle an input voltage range of about +3 to −6 volts.

6.6 BILATERAL CONSTANT-CURRENT SOURCE

The circuit in Fig. 6.19 is a two-terminal bilateral constant-current source. This circuit makes use of the channel symmetry of an FET. When terminal A is positive with respect to terminal B, the diode nearest terminal A is reverse-biased. This allows the 2-kilohm potentiometer nearest terminal A to develop a self-bias that determines the current at which the circuit will limit. The diode nearest terminal B is forward-biased, bypassing the 2-kilohm potentiometer in parallel with

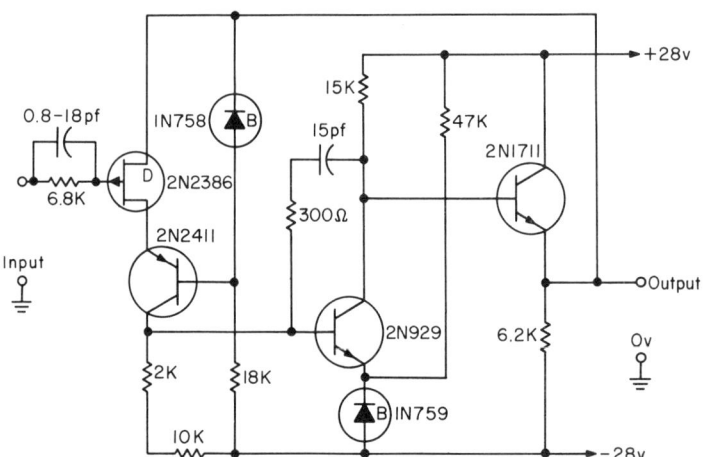

Fig. 6.18. Simple unity-gain high-input-impedance d-c amplifier.

Further Applications   115

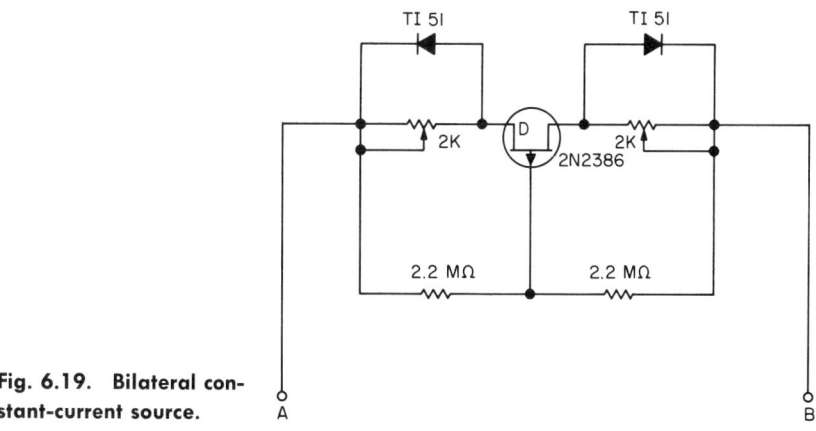

Fig. 6.19. Bilateral constant-current source.

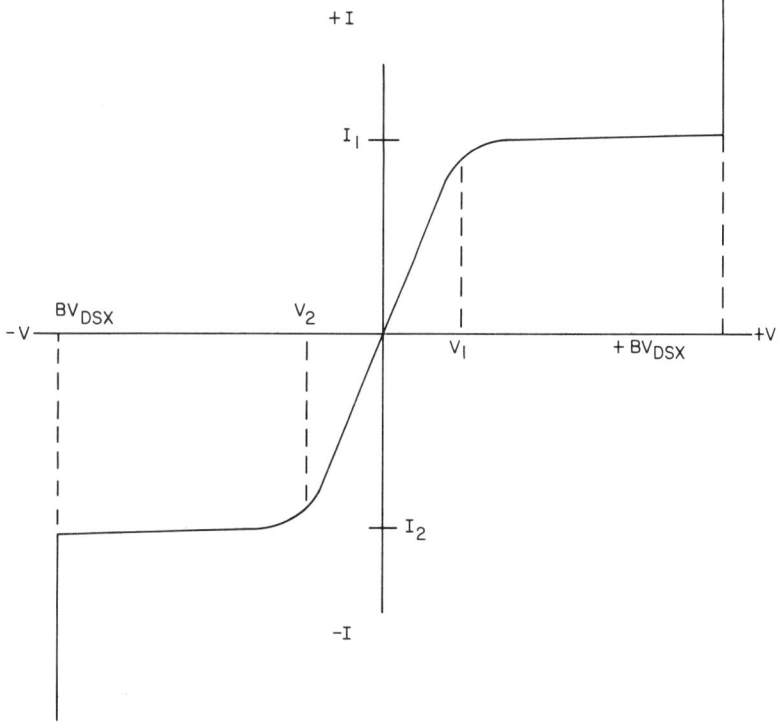

Figure 6.20

it. The voltage-current characteristic is shown in Fig. 6.20. Between voltages V_1 and V_2, the circuit behaves as a nonlinear resistor with current limiting at I_1 and I_2. I_1 and I_2 are adjusted by the two 2-kilohm potentiometers and can be made symmetrical or unsymmetrical as desired. The circuit is the dual of the double-anode voltage-regulator diode.

7

FET's in Integrated Circuits

The Integrated-circuit Movement is one of the strongest forces ever experienced by any industry. It represents a revolution of staggering proportions bringing about changes that have only begun to be felt.

Designation of this phenomenon as a Movement with a capital M is prompted by its resemblance to a revolutionary political movement. Certainly the resulting changes are and will be as dramatic as those produced by any political revolution of the past, though perhaps they will be attended by somewhat less violence. The similarity to politics seems even stronger when we consider the spectrum of philosophies in the electronics industry concerning the Integrated-circuit Movement. There are wild-eyed true believers or "Leftists" who would have us abandon every phase of electronic technology prior to the development of integrated circuits. This group is mostly confined to science fiction writers, market research managers, and co-op students. At the other extreme are the "Arch-conservatives" who want to return to the vacuum-tube era. Between these two "extremist" groups are enthusiasts and skeptics; they correspond, one would assume, to liberals and conservatives. The derivation of any further functional relationships between the electronics industry and politics is left as an exercise for the interested reader.

"Integrated circuits" is a generic term applied to several manufacturing techniques, nearly all of which fall into one of four classes.

> Circuits made up of discrete components, that are miniaturized by unusual high-density packaging techniques.
> Thin-film circuits, that have thin-film passive elements and either thin-film active elements or discrete active elements that are soldered into the circuit.
> Monolithic circuits, electronic networks that are fabricated by successive diffusions into a semiconductor substrate, usually silicon.
> Hybrid circuits, any combination of the above three. For example, it is sometimes desirable to evaporate thin-film passive elements onto a monolithic semiconductor bar into which the active and some passive elements have already been diffused. Electrical isolation is provided by a thin insulating oxide layer on the surface of the substrate.

Obvious advantages offered by integrated circuits are:

Fig. 7.1. The same PCM telemetry encoder in two versions: right, the discrete-components model; left, its integrated-circuits equivalent.

Reduced size and weight. In some cases the reduction is quite spectacular; witness the "before and after" situation presented in Fig. 7.1. Size reduction is 100 to 1 and weight reduction is 50 to 1.

Improved reliability. Entire networks in monolithic form have proved to be as reliable as single discrete components.

Low cost. This advantage is just now being realized, but in the future it will be the biggest advantage of all.

Integrated circuits still pose many unsolved problems. Probably the most important is the problem of interconnections. As higher packaging densities permit the development of more complex systems, the increase in required electrical interconnections bites deeply into the reliability advantage. Considerable research effort is now being expended on the possibility of signal transmission by "nongalvanic" means, i.e., without physical electrical connections. The most practical alternative to galvanic connections thus far advanced is optical or photon coupling.[1,2]

Most semiconductors exhibit radiative and photoconductive or photovoltaic properties. As a specific example, a gallium arsenide PN junction will emit infrared light that is a function of forward-bias current. Shining infrared light on a silicon PN junction will cause the junction to forward-bias itself (solar-cell action) and be capable of delivering an intensity-dependent current to an external load. In this way, a signal can be coupled through an optical path, from the photoemitter to the photodetector. The efficiency of this transmission link is, in general, quite low, and practical photodiodes for integrated circuits are able to deliver very little current. It follows that optical coupling is one scheme in which FET's with their high power gains and high input impedances will offer the best practical solution as the active element. The simple photon-coupled switch in Fig. 7.2 is offered as an example of the form such integrated circuits may take.

118 Field-effect Transistors

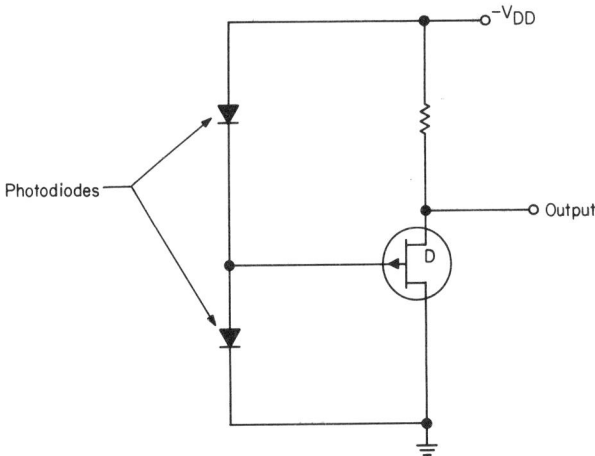

Fig. 7.2. Photon-coupled switch.

7.1 UNIPOLAR FET'S

Unipolar FET's offer at least two advantages over bipolar transistors in monolithic integrated circuits:

They take considerably less bias power. Because of this, FET's are desirable for micropower circuits, i.e., circuits that take very little power from their power supplies. The resulting low heat dissipation permits very high packaging density.

Their much higher power gain (at least at low frequencies) means that fewer devices are required to perform a given function.

A triple-diffused FET structure that has been fabricated in monolithic integrated circuit form is shown in Fig. 7.3. An N-channel structure is desirable because, for a given size and impurity concentration in the channel, the N-channel FET will have about three times the transconductance of a P-channel FET. The reason for this is that the mobility of electrons in silicon is about three times the mobility of holes.

It is desirable that these FET's operate at as low currents as possible if they are to be used in micropower circuits. These circuits are most commonly found in systems operating on low supply voltages—usually 6 volts or less—where little

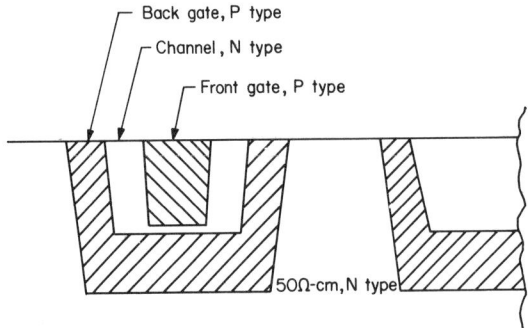

Fig. 7.3. Isolated diffused FET structure.

voltage can be spared to bias a high-current FET down to low currents. The I_{DSS} of the FET should be designed to be slightly higher than the desired bias current, so that the self-biasing drop across any source resistor will be kept to a minimum. A low-current FET should also be designed for a low pinch-off voltage to keep the transconductance as high as possible.

As a matter of fact, when an FET amplifier is to be operated with low supply voltages, the available voltage gain is determined strictly by the relationship between the supply voltage and the FET pinch-off voltage. This statement is easily proved; the gain of a common-source amplifier with arbitrary source and load resistors as shown in Fig. 7.4 is

$$A = \frac{g_m R_L}{1 + g_m R_s} \quad (7.1)$$

Since there is no external fixed bias in Fig. 7.4, the source resistor is determined by

$$R_s = \frac{V_Q}{I_Q} \quad (7.2)$$

The drain load resistor (here we are assuming that the loading of the next stage will be light) is determined by

$$R_L = \frac{V_{DD} - V_{DS} - V_Q}{I_Q} \quad (7.3)$$

From Eq. (1.54), the transconductance in terms of static FET parameters is

$$g_m = \frac{2 I_{DSS}}{V_P} \left(\frac{V_Q}{V_P} - 1 \right) \quad (7.4)$$

Now, substituting Eqs. (7.2), (7.3), and (7.4) back into Eq. (7.1) and simplifying, we find

$$A = 2 \left(\frac{V_{DD} - V_{DS} - V_Q}{V_Q + V_P} \right) \quad (7.5)$$

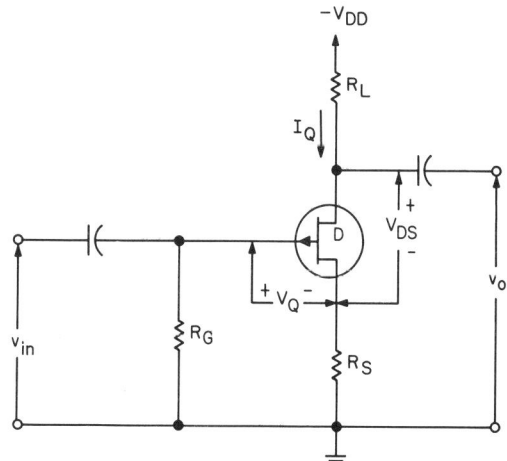

Fig. 7.4. FET amplifier stage.

Although the absolute lower limit on V_{DS} in Eq. (7.5) is obviously V_P, it is naturally desirable to keep V_{DS} larger than that. Nevertheless, the largest possible small-signal gain an FET amplifier can have is

$$A_{max} = 2\left(\frac{V_{DD}}{V_Q + V_P} - 1\right) \quad (7.6)$$

The right-hand side of Eq. (7.6) must be evaluated under worst-case conditions, i.e., minimum possible supply voltage and maximum values of V_Q and V_P. Quite obviously we would like to have both V_Q and V_P as small as possible. As was shown in Chap. 2, FET's with low pinch-off voltages have high positive temperature coefficients of both drain current and transconductance; this is borne out by the curves in Fig. 7.5. The maximum pinch-off voltage of the device occurs at the highest temperature; in Fig. 7.5, maximum pinch-off is approximately 0.5 volt.

If a 6-volt supply is available, and if there is no source degeneration ($V_Q = 0$), the maximum possible voltage gain that can be realized is 18. This is a respectable figure at such low power levels, but see what happens if the temperature drops to $-55°C$. At this temperature the drain bias current has dropped to 1.8 μa and the bias-point transconductance is

$$g_{max} = \frac{2I_{DSS}}{V_P} = \frac{2 \times 1.8 \times 10^{-6}}{0.25} = 14.4 \; \mu\mho$$

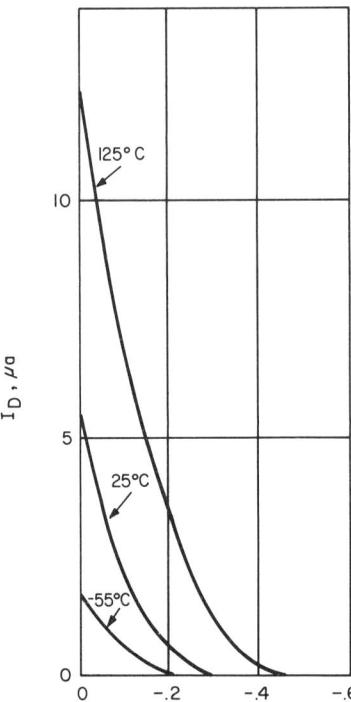

Fig. 7.5. Drain transfer characteristics vs. temperature.

The drain load resistor, determined by the high temperature conditions, is

$$R_L = \frac{V_{DD} - V_P}{I_{DSS}} = \frac{6 - 0.5}{13 \times 10^{-6}} \cong 400 \; K$$

Therefore the gain at $-55°C$ is

$$A_{min} = 14.4 \times 10^{-6} \times 0.4 \times 10^6 \cong 5.8$$

This represents a drastic decrease in gain. The conclusion is obvious: if a unipolar FET is to be designed into a micropower amplifier circuit, one must be prepared to accept wide variations in gain, unless the circuit is to be applied where the temperature range is rather restricted.

An example of an FET amplifier design in monolithic integrated form is shown in Figs. 7.6 and 7.7. Figure 7.6 is the schematic and Fig. 7.7 is an outline drawing of the circuit layout. The dimension scales in Fig. 7.7 are in thousandths of an inch. No interconnections between the individual components are shown in Fig. 7.7; these connections must be made over the surface of the silicon bar, by a process known as "stitch bonding." The capaci-

FET's in Integrated Circuits 121

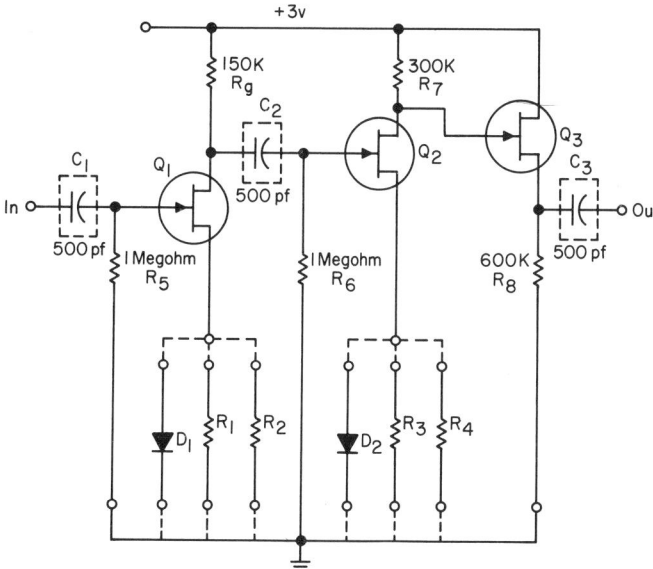

Fig. 7.6. Integrated FET amplifier.

tors C_1, C_2, and C_3 are thin-film capacitors deposited on a separate substrate of approximately the same physical size. This integrated amplifier then will require two sections of silicon bar, but the capacitors will still be of much smaller physical size than any discrete capacitors.

The gate, source, and drain contacts of the FET's are the blacked-out areas of Q_1, Q_2, and Q_3 in Fig. 7.7. Since one gate and three separate channels make up

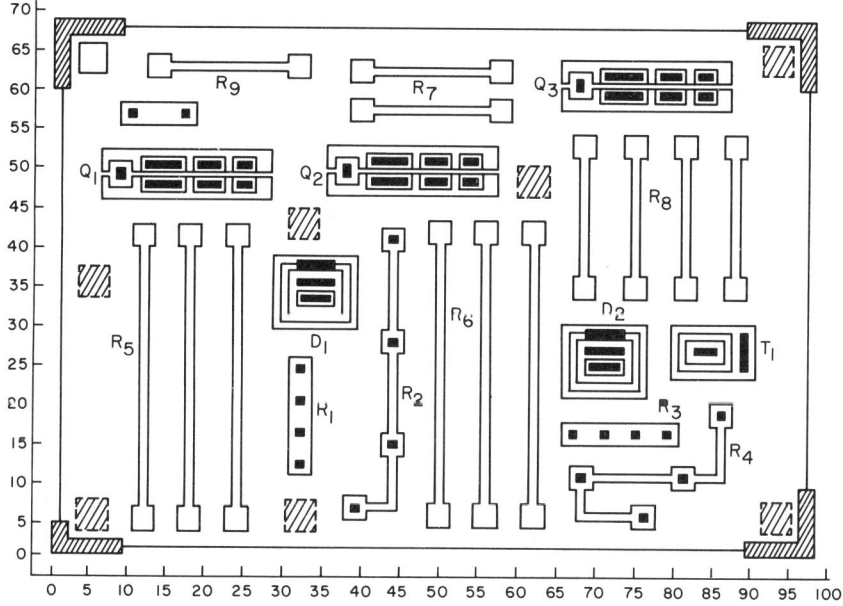

Fig. 7.7. FET network layout.

122   **Field-effect Transistors**

each FET, each FET can exhibit six distinct sets of electrical characteristics. There are also six possible combinations of the source bias resistor, and two of the resistors can be connected to any of several values. The third source resistor is nonlinear (a diode). Such flexibility significantly improves yields but requires that each circuit be probed separately to determine necessary interconnections.

This discussion has suggested some fundamental differences between design techniques for discrete circuits and those for integrated circuits. The designer of integrated circuits designs both the circuits and the devices he uses. On the other hand, the designer of discrete circuits rarely exerts direct control on device design.

Unipolar FET's are also useful in integrated logic gates where very large fan-in and fan-out figures of merit are required, because of the negligible power needed to switch FET's. Fan-in is the number of inputs a logic gate can have and fan-out is the number of like inputs that the gate will drive. Simple examples of FET logic gates are the three-input NOT AND gate in Fig. 7.8a and the three-input NOT OR gate in Fig. 7.8b. Obviously, either gate can operate with much larger numbers of inputs than are shown in the examples.

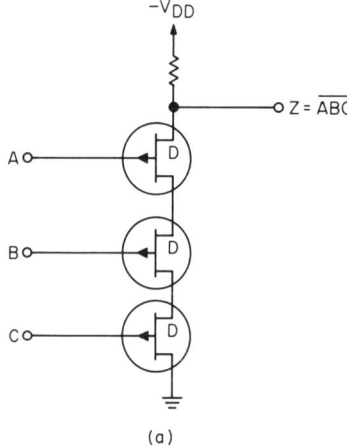

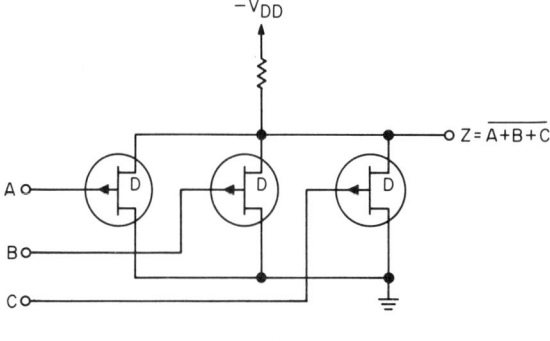

Fig. 7.8. FET logic gates.

7.2 SURFACE FET'S

A serious disadvantage of using unipolar FET's in logic circuits is the need for more than one supply voltage. This arises because the operating gate-to-source voltage is opposite in polarity to the drain-to-source voltage. Enhancement-mode surface FET's do not suffer this disadvantage. Their gate leakage currents are also several orders of magnitude less than those of unipolar FET's and are not subject to the wild temperature variations. For these reasons, surface FET's are much more desirable for integrated logic than unipolar FET's, provided that the temperature instability of the drain characteristics discussed in Chap. 1 can be overcome. Considerable effort is being expended in the semiconductor industry to solve this problem. As of this writing, however, there is no outstanding success to report. No manufacturer is yet able to mass-produce surface FET's of either type, either in integrated or discrete form.

Surface FET's can be grouped under two general classes, MOS (metal-oxide-semiconductor) FET's, and thin-film FET's. The electrical characteristics of these two types are quite similar but fabrication techniques are entirely different. The MOS structure shown in Fig. 7.9 is fundamentally a monolithic-type device. A photomask, photoresist process is used to diffuse two concentric N regions into a lightly doped P-type substrate. The oxide dielectric over the ring-shaped gate region is thermally grown at high temperatures and is about 1,000 Å thick. The metallic contacts to the source, gate, and drain are formed by evaporating a good

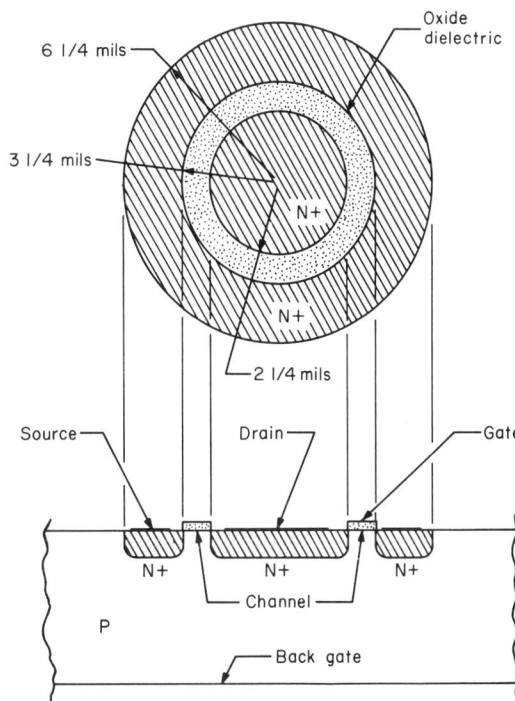

Fig. 7.9. MOS field-effect transistor.

124 Field-effect Transistors

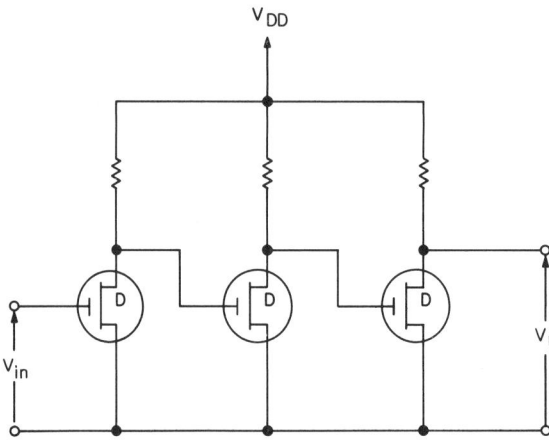

Fig. 7.10. Direct-coupled amplifier with enhancement-mode MOS FET's.

conductor such as silver or gold onto the surface. This circular geometry is one of two commonly used; the other geometry is a linear structure.

We have seen in Chap. 1 that it is possible to fabricate MOS FET's as either enhancement-mode or depletion-mode devices. An enhancement-mode device that requires a turn-on voltage of the same polarity as the supply voltage is very desira-

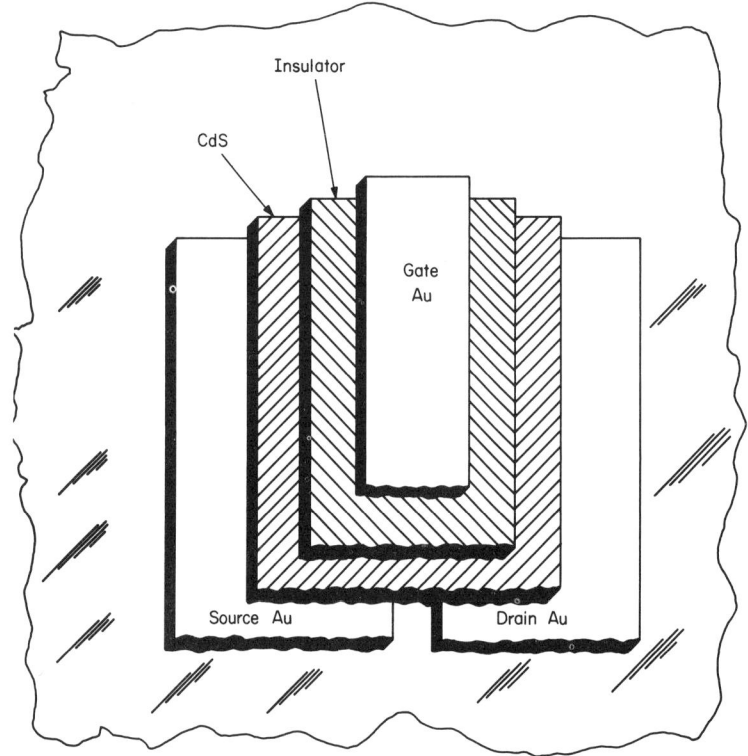

Fig. 7.11. Thin-film FET.

ble for integrated logic gates, since no reverse bias is required to turn off the FET's; an entire logic array can be operated from one supply.

A linear amplifier circuit using enhancement-mode MOS FET's is also an interesting possibility. It is possible that such an amplifier would take the form shown in Fig. 7.10. The amplifier is direct-coupled and the drain voltage of each stage provides the proper bias voltage for the gate of the succeeding stage.

Figure 7.11 shows the construction of a thin-film FET. The metal source and drain electrodes and the semiconductor channel are evaporated onto an insulating substrate, usually glass or ceramic. An insulating layer and a metal contact are successively evaporated on top of the channel film. The channel and gate dimensions shown in Fig. 7.11 are approximately the same as those of unipolar FET's.

This fabrication technique offers promise that the thin-film FET may provide the last word in inexpensive, high-density circuitry, especially for integrated logic arrays and adaptive systems. But many technological problems remain to be solved; this device is in an even more primitive stage than the MOS FET.

BIBLIOGRAPHY

1. Matzen, W., and R. Biard: Utilization of Electro-optical Phenomena to Perform Electronic Logic Functions, proposal to Wright-Patterson Air Force Base, Dayton, Ohio, Texas Instruments Incorporated.
2. Herzog, G.: Active Logic Elements Using Non-galvanic Modifying Inputs, *RCA Report*, Air Force Contract, AF19(628)1629.

INDEX

Abrupt junction, 7
Active channel, 19
Adaptation characteristics, 28
Adaptive element, 28
Amplification factor, voltage, 56
Amplifier, AGC, 78
  audio, 61
  cascode, 68–69
  class B, push-pull, 94
  common-drain, 51, 55–57
  common-gate, 51, 54–55
  common-source, 51–53, 119
  cutoff frequency, 62–63
  differential, 70
  direct-coupled, 70, 102–108
  distortion, 59–61
  frequency response, 63
  high-impedance, 69
  low-noise, 107–108
  medium-power, 98
  operational, 106–107
  RC-coupled, 61, 64
  self-biased, 52
  servo, 100–101
  transducer, 67
  video, 61
  voltage gain, 62
  wideband, 61
Automatic gain control, 76–79
Avalanche breakdown, 11, 13

Back gate, 14
Barrier potential, 1
Base, 5
Bias, constant-current, 65
  FET amplifiers, 57–59
  fixed, 52
  graphical method, 58
  self, 52
  temperature stable point, 70, 71, 81
  vacuum-tube amplifiers, 57
Bias dependence of y parameters, 42
Bias point drift, 70, 89
Bias stabilization of d-c amplifier, 70
  of power FET, 96
Bilateral constant current source, 114–115
Bipolar transistors and FETs combined, 64
Bistable multivibrator, 86–88

Bockemuehl, R. R., 5
Boltzmann's constant, 40
Bootstrapped source follower, 65–66
Breakdown voltage, 39–40
Bulk leakage current, 41
Bulk resistance effects on y parameters, 44
Bypass capacitor, 52, 64

Capacitance, bypass, 52, 64
  diode shunt, 56
  drain-to-gate, 54, 67
  junction, 3
  stray, 68
Capacitor, coupling, 62, 64
  series peaking, 64
  source bypass, 52
Cascode amplifier, 68, 69
Cathode, 5
Cathode follower, 55
Channel, 5
  conductivity, 8
  current density, 7, 9
  current flow, 7–11
  height, 5
  length, 5
  ON resistance, 40
  resistance, 18
  width, 5
Charge, density, 6
  depletion model, 1
  velocity, 13
Chopper-stabilized d-c amplifier, 91
Choppers, 89–92
Class B amplifier, 94
  efficiency, 95
  output power, 94
Collector, 5
Combinations of FETs and bipolar transistors, 64
Common drain, 51, 55–57
  amplifier, 55
Common gate, 51, 53–55
  amplifier, 51, 54–55
  equivalent circuit, 55
  input admittance, 55
  voltage gain, 54
Common source 51–53
  amplifier, 51–53, 119
  equivalent circuit, 53

127

128 Index

Common source, input admittance, 53–54
    voltage gain, 55
Commutators, 89
Condenser microphone preamplifier, 74–75
Constant-current source, 10
Constant velocity, 14
Contact potential, 1, 10, 21, 36
Current continuity, 8
Cutoff frequency, 20
    of n-stage amplifier, 63
    of single-stage amplifier, 62–63

Darlington pair, 87
Depletion layer, 1
Depletion mode, 26
Dielectric breakdown voltage, 26
Dielectric constant, 6
Difference amplifier, 70
    drift equivalent circuit, 70
Diffusion profiles, 24
Dipole moment, 27
Direct-coupled amplifiers, 70, 102–108
    chopper-stabilized, 91
    stable biasing, 70
Distortion, effect of unbypassed source resistance, 61
    harmonic, 59
    intermodulation, 60
Double-diffused FET, 24, 42
Drain, 5
    bootstrapping of, 67
Drain current, 17
    temperature dependence, 36–39
    zero temperature coefficient, 38
Drain-to-gate breakdown voltage, 39
Drain-to-source breakdown voltage, 39
Drain-to-source resistance, 32
Drift compensation, 70, 81
Drift mobility, 3, 13
Drift velocity, 13
Dynamic characteristics, 41

Ebers and Moll, 5
Electrets, 27
Electronic charge, 3
Electronic voltmeters, 108–110
Emitter, 5
Emitter-follower, 55
Enhancement mode, 27
Equivalent circuit, common-gate, 55
    common-source, 54
        with capacitances, 54
    difference amplifier, 70
    FET switch, 85
    source follower, 56
Equivalent input noise current, 49
Equivalent input noise voltage, 48
Expanded contacts, 43

Ferroelectric crystal, 27
FET squaring circuit, 80–84
    characteristics, 82
Filter, third-order active, 103
Fixed bias, 52
Fixed recombination centers, 27
Flip-flop, 86–88
    OFF equation, 87
    ON equation, 87
    radiation tolerance, 86
Forward transfer characteristic, 16, 25
Free carriers, 8
Frequency compensation, 64
Frequency response, of FET amplifiers, 63, 69, 75, 104, 108
    of FET squaring circuit, 85
    of power FET, 97
Front gate, 14

Gain-bandwidth product, 63, 64, 93
Gate, 5
    bias resistor, 67
    capacitance, 17, 18
        per unit area, 17
    current, 40
    leakage current, 40, 59, 70
Gate-to-channel capacitance, 42
Gate-to-channel diode, 16
Gate-to-channel voltage, 40
Generation-recombination noise, 46
Grid, 5
Gunn, J. B., 13

Harmonic distortion, 59
Harmonic rejection of FET squarer, 84
Heiman, F. P., 27
High-field phenomena, 13
High-frequency characteristics, 44–46
Hofstein, S. R., 27
Hybrid integrated circuits, 116
Hysteresis, 27

I_{CBO}, 94
I_{DSS}, 8, 32
    temperature dependence, 32
    zero temperature coefficient, 35
Impurity atom density, 1, 3
Induced channel, 25
Inferred pinch-off voltage, 35
Initial channel, 25
Input admittance, common-gate amplifier, 55
    common-source amplifier, 53, 54
    source follower, 56
    transducer amplifier, 67
Input resistance of FET, 40
Insulated gate, 25

Index

Insulating dielectric, 25
Integrated circuits, 116–125
   amplifier, 121
   amplifier layout, 121
   interconnections, 117, 120
   operational amplifier, 72
Intermodulation distortion, 60
Intrinsic semiconductor, 25
Inversion layers, 16

Logic gates, NOT AND, 122
   NOT OR, 122
Lumped element equivalent circuit, 41

Magnetic hysteresis loop, 27
Meter rectifier, 83
Micropower circuits, 118
Miller effect, 53
Mobility temperature coefficient, 33, 34
Monolithic circuits, 116
MOS transistors, 123
Multiplexing, 90
Multivibrators, 86–89
   bistable, 86–88
   monostable, 87–89

Noise bandwidth, 46
Noise characteristics, 46
Noise corner frequency, 47
Noise factor, 49
Noise figure, 50
   of transducer amplifier, 67
Nonlinear distributed line, 41
Nonlinear resistor, 11

OFF resistance, 84, 86
Offset voltage, 90
Ohmic contacts, 5
Ohm's law, 7, 13
ON resistance, 40, 84, 86, 90
Operational amplifier, 72, 106–107
Oscillators, phase-shift, 111
   Wien-bridge, 111–113
Output characteristics, 11
Output conductance, 10
Overshoot in choppers, 90

P-channel FET, 5
Parallel-plane diode, 7
Parameters, switch, 84
Phase-shift oscillator, 111
Photon-coupled switch, 117, 118
Physical equivalent circuit of FET, 42

Pinch-off drain current, 11
Pinch-off points, 11
Pinch-off region, 11
Pinch-off voltage, 7, 17, 32
   temperature coefficient, 35
Plate, 5
Poisson's equation, 6
Power FET, 93–101
Power transistors, 93
Preamplifier, condenser microphone, 74–75

Q-point shift with temperature, 96
Quarter-squares multiplier, 84–85

RC-coupled amplifier, 61, 64
RCA, 27
Remanent polarization, 27

Sample-hold circuit, 113, 114
Saturation, drain current, 17
   line approximate, 86
   voltage, 95
Secondary breakdown, 93
Self-bias, 52
Semiconductor bar, 2
Series chopper, 90
Series peaking capacitors, 64
Servo amplifier, 100–101
Shockley, W., 7, 21, 33
Short-circuit admittance parameters, 41
Shunt chopper, 90
Source, 5
   bypass capacitor, 52
Source follower, 55, 65, 70
   equivalent circuit, 56
Space-charge layer, 1, 6
   height of, 2
   radius of, 14
Speech compressors, 79
Spike profile, 20, 21
Square-law approximation, 22, 23, 37, 42, 77
Squaring circuit, 80–84
   waveforms, 83–84
Static characteristics, 32–41
Stitch bonding, 120
Surface FET, 123
Switch parameters, 84

Telemetry encoder, 117
Temperature coefficient, of drain current, 38
   of I_{DSS}, 35
   of mobility, 33, 34
Temperature effects, on bias point, 96
   on drain current, 36–39
   on gate voltage, 39

Index

Temperature effects, on I_{DSS}, 32
  on pinch-off voltage, 35
Temperature-stable bias point, 70, 71, 81
Thermal runaway, 94
Thermal stability of power FET, 94
Thin-film FET, 125
Timer circuit, 88
  delay time, 89
Transconductance, 9, 10, 17, 42
  at zero bias, 17
Transducer amplifier, 67
Transfer characteristic, 17, 24
Transit time, 20
Triode region, 11
Triple-diffused FET structure, 118
Triple subscript notation, 32

Uniformly doped channel, 12
Unipolar FET, 5, 6
  in integrated circuits, 118–122

Vacuum tube, 5, 9
Van Der Ziel, A., 46
Velocity-limited current flow, 14–15
Video amplifiers, 61
Voltage amplification factor, 56
Voltage gain, of n-stage amplifier, 62
  of single-stage amplifier, 62

Wideband amplifiers, 61
Wien-bridge oscillator, 111–113